DATA POWER IN ACTION

Urban Data Politics in Times of Crisis

Edited by
Ola Söderström and Ayona Datta

T0256588

BRISTOL
UNIVERSITY
PRESS

First published in Great Britain in 2024 by

Bristol University Press
University of Bristol
1–9 Old Park Hill
Bristol
BS2 8BB
UK
t: +44 (0)117 374 6645
e: bup-info@bristol.ac.uk

Details of international sales and distribution partners are available at bristoluniversitypress.co.uk

British Library Cataloguing in Publication Data
A catalogue record for this book is available from the British Library

ISBN 978-1-5292-3354-4 paperback
ISBN 978-1-5292-3356-8 ePub
ISBN 978-1-5292-3355-1 ePdf

Cover design: Qube
Front cover image: A bodaboda operator processing a delivery in Nairobi,
October 2021 (photograph by Prince K. Guma)
Bristol University Press uses environmentally responsible print partners.
Printed and bound in Great Britain by CPI Group (UK) Ltd, Croydon, CR0 4YY

Contents

List of Figures

Notes on Contributors

Sarah Barns is Senior Vice Chancellor's Research Fellow at RMIT whose interests lie at the intersection of citizen technology and urban transformation. She has led over 40 community-focused citizen technology programs that reimagine digital futures through multisensory storytelling through her practice Esem Projects. She has also practised and published widely on new models of urban digital governance. Sarah's book, *Platform Urbanism*, was published in 2020 by Palgrave.

Evan Blake is completing his doctoral studies at the University of Neuchâtel, collaborating with Ola Söderström and Nancy Odendaal on South African data politics and smart cities research. Evan's doctoral research focuses on the recasting of power relations between civil society and state actors via data and knowledge practices.

Federico Caprotti is Professor of Human Geography at the University of Exeter. His main research interest is in critically engaging with urban futures. He has a particular interest in two aspects of the urban future: the 'off-grid' city in the Global South, and the trend towards high-tech, digitally integrated and platform-mediated urban 'citizenship'. His research currently focuses on these topics in several cities in China, and Cape Town, South Africa.

Shiuh-Shen Chien obtained his PhD degree from the London School of Economics and Political Science, and now is a professor in geography, environment, and development studies, co-hired by the department of geography and the international programme of climate change and sustainable development, National Taiwan University. His research interests have covered: the Global South and international development, political economy of local and regional development, critical smart eco-urbanism, volume geography in the Anthropocene context. Chien's publications can be seen in major journals of geography, urban studies, planning and development, and area studies.

Jonathan Cinnamon is Assistant Professor in the Department of Community, Culture and Global Studies at the University of British Columbia. His research examines the societal conditions that emerge as grassroots, corporate, and government actors invest in data and information technologies, based on fieldwork in Canada, the United Kingdom, and South Africa.

Ayona Datta is Professor in Human Geography at University College London. Her research interests are in postcolonial urbanism, gender citizenship, and urban futures. She is author of *The Illegal City: Space, Law and Gender in a Delhi Squatter Settlement* (2012); co-editor of *Mega-Urbanisation in the Global South: Fast Cities and New Urban Utopias of the Postcolonial State* (2017) and *Translocal Geographies: Spaces, Places, Connections* (2011). Her recent work funded by the European Research Council (2022–26) examines regional futures emerging from the intersection of urban and digital geographies in the Global South.

Prince K. Guma is a research associate at the Urban Institute, University of Sheffield. His work explores the multiple ways through which cities and infrastructure domains are constructed and reconstructed through the diffusion and uptake of new plans, ideas and technologies. It is aimed to provide a menu for new explorations, enhance our understanding of urban possibilities, and add new insights to debates on technology and urbanity in the Global South and elsewhere.

Orit Halpern is Full Professor and Chair of Digital Cultures and Societal Change at Technische Universität Dresden. Her work bridges the histories of science, computing, and cybernetics with design. She is currently working on two projects: the first is a history of automating intelligence, democracy, and decision-making; the second project examines extreme infrastructures and the history of experimentation at planetary scales in design, science, and engineering. Her current book with Robert Mitchell (MIT Press, 2023) is titled *The Smartness Mandate*.

Rob Kitchin is Professor in Maynooth University Social Sciences Institute and department of geography. He was a principal investigator on the Programmable City project (2013–18) and Building City Dashboards project (2016–20). He has published widely on the relationship between space, time, digital technologies, and society.

Nancy Odendaal is Associate Professor in City and Regional Planning in the School of Architecture, Planning and Geomatics at the University of Cape

Town. Her research and teaching focus on the role of technology in urban change, the use of technological tools in the urban planning process, and the relationship between spatial change, service delivery and infrastructure. Recent publications focus mainly on smart cities 'from the bottom up' – how new technologies and data can facilitate urban inclusion and spatial transformation towards more diverse and inclusive cities.

Alison Powell is Associate Professor in Media and Communications at the London School of Economics and the founding program director of the data and society master's programme stream. She directed the JUST AI Network, supported by the AHRC and the Ada Lovelace Institute: see https://just-ai.net. JUST AI created alternative ethical spaces, practices, and orientations towards the issue of data and AI ethics within a broad community of practice. She is the author of *Undoing Optimization: Civic Action and Smart Cities*, published by Yale University Press, and has teaching and research interests in both critical data studies and media futures.

AbdouMaliq Simone is Senior Professorial Fellow at the Urban Institute, University of Sheffield, and Visiting Professor of Urban Studies at the African Centre for Cities, University of Cape Town. Key publications include, *For the City Yet to Come: Urban Change in Four African Cities* (Duke University Press, 2004), *City Life from Jakarta to Dakar: Movements at the Crossroads* (Routledge, 2009), *Jakarta: Drawing the City Near* (University of Minnesota Press, 2014), *New Urban Worlds: Inhabiting Dissonant Times* (with Edgar Pieterse, Polity, 2017), *Improvised Lives: Rhythms of Endurance for an Urban South* (Polity, 2018), and *The Surrounds: Urban Life Within and Beyond Capture* (forthcoming, Duke University Press).

Ola Söderström is Professor of Social and Cultural Geography at the University of Neuchâtel, Switzerland. His work focuses on global dynamics of urban development, urban material culture, urban visual cultures, and tactics of urban living. His recent work has investigated critical forms of mobility (*Critical Mobilities*, Routledge, 2013), trajectories of urban globalization through relational comparisons of cities of the Global South (*Cities in Relations*, Wiley-Blackwell, 2014), smart urbanism, and relations between urban living and psychosis.

Petter Törnberg is Assistant Professor at the University of Amsterdam, focusing on the critical study of platforms and AI through digital data and natural language processing. He is particularly interested in political radicalization and polarization. Petter is currently pursuing a NWO Veni project on the production of urban place on digital platforms, studying urban representation in large-scale textual data.

Ying Xu is a postdoctoral research associate in the geography department, University of Exeter, UK. He received his PhD degree from the department of geography and resource management, The Chinese University of Hong Kong, Hong Kong S.A.R., China. He has research interests on sustainable cities, urban governance, and spatial-temporal heterogeneity of environmental issues.

Notes on Funding

This book has benefited from the generous funding of the Swiss National Science Foundation (SNSF). It draws on research funded by the project Smart Cities: 'Provincializing' the global urban age in India and South Africa (SNSF grant: 10001AM_173332) and the project Data Politics and New Regimes of Mobility and Control During and After the COVID-19 Pandemic (SNSF grant: 51NF40_205605). The latter is a project part of the SNSF – National Center of Competence in Research (NCCR) On the Move. We are also thankful for SNSF support allowing this edited volume to be open access.

Swiss National Science Foundation

Data Power in Action: Urban Data Politics in Times of Crisis

Ola Söderström and Ayona Datta

Introduction: why data?

While working on this book, one of the editors had to handle the consequences of a cyberattack of his university server by Conti, an infamous Russian group of malevolent hackers. As a consequence, he had no access to all his files for weeks, and some of his personal data were accessible on the darknet. Faculty members were left to speculate why their university was targeted, what the hackers hoped to get out of this, and if a ransom was addressed to the direction of the university. This is one of many instances where our contemporary dependence on data and our related vulnerability becomes very tangible. It shows that data is everywhere, increasingly mediating and shaping all domains of life (work, leisure, kinship, friendship, sexuality).

Hacking, the term officially used to describe the above incident, seems inadequate though. This is a form of data theft in which personal data becomes the new currency of international criminal activity. Even as data flows through our handheld devices, communications towers, satellites, undersea cables, and the whole assemblage of infrastructures that make data flows possible, personal data itself provides the accumulatory capacities of capital of our current global condition – a condition that Manuel Castells had labelled as the 'informational capital' in the network age. Personal data, of course, is also subjective – it is marked by the conditions of production of our bodies in digital space. Personal data marks the onset of knowledge about people, spaces, and places, and therefore speaks to the political condition of our current moment. Whether we consider widely mediatized events, such as the role of Cambridge Analytica in

Brexit or the 2016 US elections, we are today increasingly aware not only of the power of data but, more importantly, of the diverse nature of this data, its flows through our bodies and the possibilities of its disruption. Data politics emerges at this junction as the data deluge becomes highly diversified, personalized, compartmentalized, but also fragmented, disconnected, and uneven.

Despite the current proliferation of literature on big data and data analytics, the political nature of data is often left unattended or implicit. One of the key corrections to this gap by Bigo, Isin and Ruppert (2019) notes that data politics emerges through the newly mediated relationships between the state and the citizen that generate 'new forms of power relationships and politics at interconnected scales' (Bigo et al, 2019, 4). For Bigo et al, data politics is a poststructuralist reorganization of power where the production and circulation of data produces a transformation of the relationship between technology and people at all scales of data production and circulation. Yet even though scholars increasingly argue, as Bigo et al, that data is inherently political, they are rarely explicit about the diverse and fragmented nature of data, particularly in the Global South (but see Arora, 2016; Milan and Treré, 2019). Thus, much of the investigation of data politics focuses on what concerns researchers in the Global North – big data, data infrastructures, cybersecurity, surveillance, and so on. These assume data to be political yet focus less on the disrupted, fragmented, and disconnected nature of data flows and the politics therein.

To understand this, we need to turn towards data politics in the Global South. As Bowker et al (2010, 103) argue, 'people, routines, forms, and classification systems' are integral to data infrastructures, which reorganize ethical and political values embedded in the production of data. They note that categorization and standardization lie at the heart of sorting out data, and yet these processes are embedded in historical, social, and geographical flows of power. This can be seen as Biruk (2018), in her account of 'cooking data' by survey data collectors in Malawi, highlights the labours of collection, production, circulation, and storing of data as an inherently political process. She points to the tensions between standardization and improvisation of data that highlight the ways that data is both political and politicized in the Global South. Similarly Agrawal's research on Indian censuses notes that data can act as a 'political weapon' (Agrawal and Kumar, 2020) of the state to enact a rule of law over territories and populations. In the Global South, then, data may not translate seamlessly into information as it is often bound to its invisibility, scarcity, and even disconnectedness on the flows of data across people, workers, and institutions. As Datta (2023) argues, the uneven flows of data produce 'informational peripheries' of the state – spaces where 'exclusions are marked by both geographic and informational distance from the digitalizing state'. It is in the claims to a seamlessness of data in the digital age and the practices that expose its fractures across spaces and scales that we understand data politics to emerge.

In this context, this book has three aims. First, it focuses on data politics in the urban realm, which is at the same time a terrain of deployment, resistance, regulation, and subversion of data power. Therefore, the book investigates *urban* data politics: how we see the city and its citizens through a particular set of data, how the state uses data to visualize and govern its citizens and territory, and how this data is then used by civil society and non-state actors for making the state and private actors accountable. Here data politics may also be understood through the three ways that Degen and Rose (2022) propose as the reconfiguration of the urban by digital technologies – storytelling, animation, and seamfulness. They argue that these three terms 'are ways of describing how different configurations of the new urban aesthetics are organized and put into practice.' While storytelling is a way of narrating the urban moment through the digital, animation is identified as 'emergent qualities of digital mediation'. However, it is 'seamfulness' – 'a critical term, [which] attempts to reconfigure the distribution of visibility to make that invisible labor available to perception' – which unravels urban data politics. While Degen and Rose were referring to the digital age producing a new urban aesthetic, their argument can be stretched to examine the seamfulness of labour across citizens, civil society, and state in making invisible data visible, in interpreting and obfuscating data, and in producing new stories, mediations, and labours to do the work of data.

Second, while the book is framed by a set of chapters on the role of (infra)structural trends in the world of computing (the rise of big data and algorithmic power), contemporary capitalism (the rise of data and platform capitalism), governance (the rise of sensory or complex power), and ethics (the rise of data activism and issues of data justice), it is primarily practice-orientated. Here we see the everyday as entangled with narratives of data power and governmentality, which Castells (2010) had accurately observed as 'the power of flows takes precedence over the flows of power'. Big data and algorithmic logics operationalize flows of data as a virtue, whereas Simone and Rao (2021) note:

> At best, big-data integration positions those traced as elements of a set or as data points within databases whose parameters change continuously, depending on who is viewing the data, with what other databases these individual points are being linked, and for what specific, instrumental purposes those links are being forged.

This is a practice perspective: that is, one that focuses on the instrumental purposes for which data is deployed. This approach is timely and necessary, in view of the predominance of structural accounts in urban studies, and apt to reveal the entanglements, tensions, and spaces of possibility and hope in urban data politics.

The third aim of the book is to look at crises as moments of acceleration, visibility, and legitimation of new forms of data power. We understand the currency of data to be generated by the mode of continual crises that unfolds in the city. This includes immediate crises such as the recent COVID-19 pandemic and the war in Ukraine, but also longer crises such as the slow erosion of data ethics and autonomy, the erasure of civil rights and the public domain, as well as the co-optation of the digital public sphere for profit. Several chapters deal with these multiple crises not least owing to the fact that they derive from a context and from studies conducted during the pandemic, but also because these long-term crises make visible, accelerate, and often legitimize further data colonization by global corporations and the state. Here we understand data as contingent upon the spatio-temporality of its flows through infrastructures, devices, and our bodies across different scales. Data is historical, real-time as well as speculative. Data produces particular ways of seeing the everyday and its fungibility across spaces and scales. The spatio-temporality of this data produces what Amoore (2011, 24) calls the 'data derivative' – 'a visualized risk flag or score drawn from an amalgam of disaggregated fragments of data, inferred from across the gaps between data and projected onto an array of uncertain futures'. The data derivative potential produces the narrative of crisis – for if data could predict or speculate about a future time, the time of the present could be customized to fit this desirable future. Crisis, then, presents data as a series of time narratives that link past decisions to future potentialities, present actions to future aspirations.

To understand data power in action in contemporary cities, we set the stage in what follows for how it operates today in contemporary capitalism and its variegated forms of governmentality across the Global North and South. By focusing essentially on urban situations outside Europe and North America – in Kenya, China, India, South Africa – this book provides a perspective on data politics beyond data universalism: the idea that the data deluge would unfold in the same way with same consequences everywhere. While chapters in the first part of this book highlight a series of planetary trends in the rise of data power, the more empirically based chapters in the two other parts show that data politics can only be envisaged as ontologically, ethically, and epistemologically variegated (Milan and Treré, 2019).

This introduction discusses the concepts that are central to this book – urban data politics, data power in action and crises – then moves on to explain the structure of the book and highlight the main arguments of its chapters.

Urban data politics

Data has objectivist connotations. It is a key word of positivist approaches in science, and its etymology – plural of Latin *datum*, what is given – evokes unmediated and obvious facts. Data *politics* could thus sound like an oxymoron. For the common sense, data like artefacts do not have politics. However, as soon as what we call data in research is examined, their mediated and constructed characters come to the fore. They should thus rather be called *capta* (Kitchin, 2014, 2) to remind us of the actions – selections, measurements, samplings, and so on – needed to turn the world into data. Seeing *data* as *capta* opens the possibility of data politics, where data is produced, selected, used, and contested within power struggles.

Data as used in public life, rather than the scientific arena, is historically related to the state and the emergence and development of statistics. The genealogy of modern statistics intertwines three different threads: the English 'political arithmetics', based in particular since the 17th century on parish registers; the German *Statistik* with its roots in the 17th century, which aims to develop a comprehensive and descriptive understanding of a human community; and the French centralized administration's practice of using data for government since the 18th century (Desrosières, 1998). Together, these three practices were the source of the national statistical offices created in Europe in the 19th century.

These practices are developed for the state and by the state, which across these centuries and until recently held a quasi-monopoly over data regarding human populations and the characteristics of their lives. During the past 30 years, with the development of the world wide web in the 1990s and digital platforms in the 2000s, this quasi-monopoly has been seriously eroded: 'the sovereignty of the state in accumulating and producing data about its population, territory, health, wealth, and security is being challenged by corporations, agencies, authorities, and organizations that are producing myriad data about subjects whose interactions, transactions, and movements traverse borders of states in new and complicated patterns' (Ruppert et al, 2017, 4).

As a consequence, actors that were marginal until the late 19th century – private corporations, civil society organizations, and citizens – have come to play a more important role in the production, analysis, and circulation of data concerning populations and the world at large. This data production concerns phenomena such as consumer preferences, emotions, patterns of mobility, access to services: data that was and is often not produced by the statistical registers of the state. This 'non-state data' is also accumulated for different purposes, from profit-making through the monetization of data sets by digital platforms to data-based rights claims, rather than taxation and biopolitical control.

5

These new power geometries in data production go hand in glove with the rise of data as a central aspect of cultural, economic, and political power. Data has become a central mediation in social life, from everyday cultures (Burgess et al, 2022) to contemporary regimes of governmentality (Isin and Ruppert, 2020) through capital accumulation (Barns, 2020). It can be argued that datafication – that is, 'the transformation of social action into online quantified data' (van Dijck, 2014, 198) – is today as central to social change and order as mechanization and electrification were in the past (Couldry and Hepp, 2016). Therefore, data is much more than anecdotal in contemporary politics, and in particular in urban politics: it is deeply inscribed in its mechanisms.

While we acknowledge that much of the work on data so far has focused on the production and social shaping of data (Kitchin, 2014), the racialization and gendering of algorithms (Noble, 2018; Strengers and Kennedy, 2020), as well as the uneven geographies of digital infrastructures across different scales (Furlong, 2020; Guma, 2020; Datta, 2023), in this book we focus specifically on the political potential of *urban* data – in ways that it is both weaponized and democratized in urban contexts. This is a relatively new and topical theme in the context of current crises that are emerging in cities across the Global South in particular. While there has been much focus on smart cities and digital urbanism (Söderström et al, 2014; Barns, 2018; Datta, 2019; Guma and Monstadt, 2021), this book brings together the two themes of data and urbanism through their power geometries and political confluences. It brings together critical geographies of the urban in conversation with the political geographies of data to argue that urban data power is much more than smart cities or platform capitalism. Urban data power is both a continuation and disruption of historical power asymmetries, from within and beyond the state, in partnership and disruption of global corporations, co-constructed and in parallel with civil society actors.

While the majority of work on digitalization and the city focuses on digital infrastructures or algorithmic power, how they are politically, socially, and economically shaped, this edited collection focuses on the power of data in action. Since the pioneering work of Graham and Marvin (1996; 2001), the digitalization of the city has been approached in urban studies primarily as a networked infrastructure reworking the organization of cities, introducing new forms of inequalities in terms of access, autonomy, and rights. The important work on platform capitalism (Srnicek, 2016; Zuboff, 2019) adopts a similar (infra)structural viewpoint. Agency within these urban digital infrastructures, both of the powerful and the less powerful, has been given less attention. In contrast, agency with and through digital devices has been central in media studies (see, for instance, Milan, 2013; Couldry and Hepp, 2016; Stephansen and Treré, 2019; Burgess et al, 2022) but rarely focused on cities. The specificity of this edited book is to bring an agency perspective to our understanding of the digitalization of cities.

Data power in action

As Kennedy and van Dijck argued a few years ago: 'Thinking about agency is fundamental to thinking about the distribution of data power. And yet, in the context of datafication, questions about agency have been overshadowed by a focus on oppressive technocommercial strategies like data mining' (Kennedy et al, 2015, 2). Since then, there has been a response to their call and also to calls by others (for instance, Couldry and Powell, 2014) in the field of (critical) data studies to balance the famous structure–agency scale. Agency is, of course, a broad and multifaceted category: it covers a broad range of practices, from state officials or corporate CEOs taking decisions on data collection and analysis, to 'click workers' employed by AI firms (Casilli, 2017), Uber drivers (Attoh et al, 2019; Pollio, 2019) or data activists (Milan and van der Velden, 2016). Only part of this array of actors and practices has been covered and unevenly across disciplines. Most of this work has been on data activism (for instance, Beraldo and Milan, 2019; Milan and Treré, 2019); less has been done on ordinary everyday data practices (but see Lupton, 2018; Burgess et al, 2022) and, as previously mentioned, mostly in the field of media and cultural studies. In contrast, issues of agency have been a minor melody in urban data studies. Front stage has been occupied by important work on social sorting through technology (Graham, 2005), the corporatization of urban governance (Hollands, 2008; Söderström et al, 2014), the critique of techno-utopianism (Datta, 2015) or platform urbanism (Barns, 2020). However, the number of exceptions to the rule of work centered on data infrastructures has been growing in recent years with studies of citizen sensing (Gabrys, 2014; Houston et al, 2019), data-based activism and its limits (Cinnamon, 2020; Chapter 10, this volume) or platform workers (Attoh et al., 2019; Pollio, 2019).

We agree that more needs to be done and that studying data power and data politics requires to move beyond important and necessary critical work on the power of the extractive and surveillant logics of digital platforms (Zuboff, 2019), because 'condemning surveillance is not the whole story of our datafied times' (Kennedy et al, 2015, 1). It is necessary to not simply rehearse the 'Big Critique' and its 'tendency to mirror the rhetoric of Big Tech', reinforcing the claims it makes about itself (Burgess et al, 2022, 13–14) and thus producing an incomplete picture of the power of data in contemporary societies. We need to better understand the daily uses of data in projects such as data clubs (Powell, Chapter 4, this volume), listening to the voice and narratives (Couldry and Powell, 2014) of users such as delivery workers (Guma, Chapter 9, this volume), civil society organization leaders

(Blake et al, Chapter 11, this volume) or international consultants and leaders of national data strategies (Datta and Söderström, Chapter 7, this volume).

Yet, focus on agency and its autonomy does not have to be separated from critical neo-Marxist (Thatcher et al, 2016) or neo-Foucauldian (Isin and Ruppert, 2020) perspectives. On the contrary, we should strive to study agency in the context of and in tension with renewed strategies of accumulation and regimes of power, much as, a long time ago, de Certeau (1984) described, in the introductory pages to his *grande oeuvre,* his street-level approach to everyday life as a complement to a Foucauldian view from above the street. Remembering those classic pages, we have used de Certeau's famous dialectic couple 'strategy/tactic' to organize the chapters of this book, as we develop below.

We concretely address the question of agency in the book by focusing, on the one hand, on the generative role of data in the urban world and, on the other hand, on its everyday use, rather than its logic of production. First, on a general level, we take from Bigo et al (2019) the idea that investigating data politics is investigating data as generative in the political order of the city. It is generative as ideology, in the form of *dataism* – the 'belief in the objective quantification and potential tracking of all kinds of human behavior and sociality through on-line media technologies' (van Dijck, 2014, 198) – or 'data positivism', for which 'whatever aspects of the social are not digitally captured are relegated to non-knowledge' (Power, 2022, 11). This generative ideology is, for instance, at work in the imaginaries of the Smart City Mission's officials in India (Datta and Söderström, Chapter 7, this volume). It infuses a 'data epistemology', a way of seeing and cognitively organizing the urban world as made of 'clusters and patterns, located within a larger data structure' (Törnberg, Chapter 3, this volume). This way of seeing is in turn generative of social ordering practices (Couldry and Hepp, 2016), where people are assigned to clusters through which acts of governing are performed (Isin and Ruppert, 2020). The sorting of good and bad citizens in the Chinese Social Credit System, which Xu et al (Chapter 8, this volume) describe as variegated rather than totally unified, is emblematic of these practices of social ordering that governments tend to hand over to private data analytics companies, like Palantir (Powell, Chapter 4, this volume).

These examples are related to practices of datafication that citizens are subject to or resist. Agency resides also, beyond resistance to datafication, in the production and tactical use of data: *data-making,* 'a strategic mode of agency that can arise if the subjects of datafication are given tools to both understand and work with the data that they produce' (Pybus et al, 2015, 3). Several chapters in this book investigate the practices, limits, and possibilities of using data as a tactic tool for progressive urban politics both in the Global North (Powell, Chapter 4; Barns, Chapter 6) and in the Global South (Cinnamon, Chapter 10; Blake et al, Chapter 11). If this volume looks at data power in action both from the perspective of the powerful and of everyday practice, it also strives to be more than critical, reflecting in each chapter the possibilities of progressive data politics.

Crises

Finally, this book focuses on how data politics and data power play out in times of crisis: how data politics are shaped by crises and their narratives and how data shape crises. This is in part due to the general sense that our present time is characterized by a series of deep crises affecting various forms of futures: the planetary with global warming; the biopolitical in the broad sense (from the mass extinction of species to the multiplication of pandemics); the geopolitical with, notably, the war in Ukraine. These are big crises that shape urban data politics in different ways, but there are also slow-burning everyday crises, for instance in the provision of housing and basic urban services. Chapters in this book engage with this broad array of crises as critical junctures, moments of inflection in data power and politics. They also provincialize the common-sense idea in the Global North that crises are sudden and unexpected events, by investigating urban situations where crises are the normal condition of everyday life for a majority of the population.

Envisaging crisis as a characteristic of an epoch and an interpretive frame is, of course, itself a *topos*, a mode of thinking which has deep roots in the history of modernity (Koselleck and Richter, 2006). Thinking of crisis as a central driving force of social change is the hallmark of a structural or systemic view of society where, for instance, in a Marxist tradition the contradictions of capitalism inevitably lead to crises, themselves working as forces of social transformation. In a period when the predictions about a coming major crisis of capitalism by observers of its *longue durée* (Wallerstein et al, 2013) seem to materialize, we are inclined to foreground this topos once again.

However, there is a more specific reason why crisis has a particular resonance when analysing data politics. Crisis is not only a topos of social theory, but it surfaces constantly as an emic category in the reflections and actions of actors on the ground. It acts as a major form of evaluation and justification (Boltanski and Thévenot, 2006), for instance when discussing why war rooms are required to harness data in the Indian COVID-19 management strategy (Datta and Söderström, Chapter 7, this volume). This is because crisis is a central element in the 'smartness mandate'; that is, smartness not as a recent marketing strategy of IBM or Cisco, but as an epistemology with its roots in technologies and scientific theories of the 20th century Cold War period (Halpern and Mitchell, 2023). In this form of thinking, which structures smart city and other techno-solutionist narratives today, resolving crises through computational strategies of resilience is the central justification of the deployment of sensor-, data- and algorithm-intensive systems of intervention. Shocks and crises are opportunities for their deployment. Rather than inviting an inquiry into (and action on) their causes, crises in this epistemology are framed as inevitable problems that should be accepted and mitigated. This crisis/resilience model derives according to Halpern and Mitchell (2023, chapter 4) from the merging of ecological thinking about resilience in the 1970s with business practices becoming progressively a 'new normal'. In other words, we focus on crises, because it is a cognitive register closely enmeshed with data politics in its mainstream form, ubiquitous in the words and actions of the economic and political elite. How this data-powered resilience strategy encounters the everyday grapplings with the banality of urban livability crises constitutes one of the questions of this book (Simone, Chapter 5; Guma, Chapter 9; Cinnamon, Chapter 10; Blake et al, Chapter 11, in particular). But we provide no clear-cut answer to this question. We rather invite readers to avoid dichotomies, such as the ones that derive from a superficial reading of Lefèbvre (1991) (where 'representations of space' are pitted against 'spaces of representation') or de Certeau (1984) (tactics against strategies). We rather suggest that it is productive to pay attention to the homologies between everyday and technologically sophisticated practices of computation (Simone, Chapter 5, this volume) and imagine hybrid forms between phenomenology and data sciences, such as explored in data feminism (D'Ignazio and Klein, 2020).

Structure and contents of the book

While the book focuses primarily on data practices, chapters in this book envisage them in constant relation with material infrastructures, governance structures, and mechanisms of data capitalism. All contributions try also to avoid a simple domination/resistance framework. Chapters emphasize relations between logics and actors: mediations (Couldry and Hepp, 2016; Degen and Rose, 2022), glitches (Leszczynski, 2019; Leszczynski and Elwood, 2022), and resources of hope (Burgess, 2022) rather than the irresistible unfolding of a single logic of data power.

To set this moving stage, Part I of the book, entitled 'Frames', looks at broad (infra)structural trends or questions common to the more specific issues discussed in the two following parts on actors' strategies and tactics. To observe data power in action, we then distinguish (classically) between strategies and tactics using the well-known, but nonetheless useful, opposition elaborated by de Certeau (1984) where strategies are related to institutions and a durable system of rules, while tactics are practices developed in specific temporal and spatial situations, searching for leeway and trying to circumvent these fixed rules. Strategies and tactics are thus intertwined, and it is a choice of perspective to focus primarily on the first or the second.

Thus, in Part II of the book, contributions focusing on *strategies* look at the role of cities that try to develop alternatives to the power of platforms, and in contrast at authoritarian states, such as China and India, shaping or reshaping their 'technopolitics' in times of crisis. In Part III, contributions focus on the *tactics* of civil society organizations, delivery platform workers or cities that try to use the interstices of state-led smart city policies in South Africa and Kenya.

In the opening Chapter 2 of Part I, Rob Kitchin draws on his long-standing engagement with data politics to depict a broad picture of the structural processes at play. Kitchin argues that, while there is a long history to data power, big data, on which his chapter focuses, has significantly increased the power of data 'to maintain control or extract profit, or to socially sort people along the lines of race, ethnicity, gender, class, sexuality, disability and other social markers'. Data is today central to a new phase of capital accumulation, data capitalism, where smart cities foster a market-orientated approach to urban governance and digital platforms colonize everyday urban life to extract and monetize data. These transformations – through which already existing inequalities and exclusions are amplified, citizens are recast as consumers, and surveillance becomes ubiquitous – profoundly reshape, Kitchin argues, governmentality and pose important questions in terms of social justice and democracy. Data ethics, data justice, and data activism initiatives, discussed in their different forms in the last part of his chapter, are responses and forms of resistance to data power. In this context, we cannot produce a single narrative about unfolding data politics but should view it as a relational process, allowing some hope in increased democratic control.

In Chapter 3, Petter Törnberg examines platformization as the rise of governance through data power. He approaches platformization, born through the 2008 financial crisis, as a form of accumulation based on the privatization of employment regulation and as a 'way of seeing': an epistemology. The power of platforms rests, Törnberg argues, on the constitution of proprietary markets, which have complemented previous national or transnational markets, controlled by private transnational companies through digital technology. While, like neoliberalization, platformization is variegated, it shares a series of characteristics: technosolutionism, the attempt to impose its own market rules, and a shift of responsibilities onto their users. Platform power also consists in the emergence and imposition of an epistemology for which the world consists of 'clusters and patterns, located within a larger data structure'. This epistemology drives a form of governance characterized by technoliberalism (Malaby, 2009) and its 'trust in the invisible hand of the platform algorithm'. In front of this unhinged data power, Törnberg concludes, data must become the object of democratic regulation and control.

In Chapter 4, Alison Powell investigates ethics as a practice in data-driven contexts. Powell argues that scale and interscalar connections are crucial in this respect as 'many aspects of the current ongoing crisis [notably the climate crisis] are experienced at small or lived experiential scales through bodily perception, while only being able to be experienced at a global or distributed scale through data and the narratives created based on it'. Powell discusses data ethics and justice in the context of the COVID-19 crisis in the UK where large-scale data analytics have been, as elsewhere, delegated to private companies such as Palantir, whose practices had problematic unethical biases. Powell's response to these issues is not a celebration of small scale. She rather stimulates the imagination of 'other possible futures' by discussing practices of data commoning and data-sociality (as there is bio-sociality around illness and diagnosis) in projects in Bristol and London.

In Chapter 5, AbdouMaliq Simone provincializes the narratives of data capitalism by focusing on the urban majority in cities of the Global South caught between a data apparatus of surveillance and extraction, and a different ontology of data as information and knowledge that can be made operable to navigate the uncertainties and complexities of everyday life. Drawing on Hui (2016), he defines data as 'not a discrete object as much as a mode of existence to be enfolded into a decision, legitimation, or prediction'. He asks how data are produced and used in such uncertain situations, contrasting with the supposedly increasingly predictable and transparent urban world of the digital age. However, rather than opposing the logics of data capitalism and the everyday data practices in the Global South, he points to homologies in their operations: how, for instance, the Kebayoran Lama market in Jakarta works as a sophisticated interoperable data infrastructure. Therefore, Simone argues, if we want to fully understand urban data power, we need to conceptualize data 'beyond conventional modes of calculation, measurement, and value'.

Part II explores strategies in the landscape of urban data politics. This does not mean that chapters focus simply on powerful actors but on practices and initiatives that are characterized by forms of planning, institutionalization, and rules.

In Chapter 6, Sarah Barns discusses how municipal reform practices are responding to issues of data access and availability in reaction to the rise of platform services across cities. The imperative to 'take back our data' is, she argues, no longer confined to a radical fringe, but is reflected in collaborative agendas being pursued by governments, civil society, and industry at municipal, regional, national, and supranational scales. In this emergent landscape, cities can play a key role by developing novel approaches to data governance that defend the rights of citizens from wider platform practices of data accumulation and surveillance. To explore these city-scale alternate modes of data politics, Barns contrasts Barcelona's digital reform programme, Toronto's 'civic data trust' concept, and the initiatives of the Cities Coalition for Digital Rights to support municipal data governance.

She argues that these experiments are a testimony of the enduring vitality of cities as sites of struggle and agency in the digital age.

In Chapter 7, Ayona Datta and Ola Söderström focus on 'COVID War Rooms' created through a repurposing of the control and command centres of Indian smart cities in the context of the coronavirus pandemic by the Indian Smart City Mission. They study this process in the making in webinars that took place in the early days of the pandemic. Datta and Söderström analyse these webinars and war rooms as sites of data- and technopolitics in the making, where the pandemic works as test bed, accelerator, and legitimation for the full use of the smart city's surveillant affordances. They argue that these are sites where 'smartness as epistemology' can be observed at work. However, they conclude, Indian urban data politics, both highly centralized in its organization and much more fragmented across scales and actors when observed in action, blurs the idea of a frictionless roll-out of this way of seeing and organizing the digitalized city.

In Chapter 8, Ying Xu, Federico Caprotti, and Shiuh-Shen Chien deal with China's Social Credit System (SCS), initiated in 2014 and determined by fears of impending and potential crises. The SCS is, in their view, an example of the evolution of smart into platform urbanism with intermediation as its main function. The core of this intermediation is the top-down shaping of citizenship. Based on a governance mode focused on 'smartmentality' (Vanolo, 2014), the SCS is instituting a new or at least revised moral order in urban life, by introducing specific technical parameters and behavioural codes in order to distinguish between 'good' and 'bad' citizens. The chapter analyses the SCS discourse proposed by the national government, as well as different types of municipal SCSes adopted across two Chinese cities (Hangzhou and Tianjin). The chapter explores in depth how the SCS is operationalized, as well as the role of market actors and urban residents. Xu et al show that rather than a centralized, nationally uniform and fully connected system, as it is often portrayed in the media, the SCS is, as yet, a municipally diverse set of emerging practices orientated by differing conceptions of 'the good citizen'.

Part III of the book looks at more interstitial practices, ways of doing with or improvising within frameworks, processes, and rules set by economic or political institutions: de Certeau's 'tactics'.

In Chapter 9, Prince K. Guma focuses on digital platforms in Nairobi to examine articulations of platform work, everyday life, and survival in times of crisis. He offers a postcolonial critique on precarious work through ethnographic stories of how at the height of COVID-related socio-spatial inequalities, residents appropriate different digital systems and delivery platforms to navigate urban problems and restrictions. Guma also demonstrates how, while filling certain voids during COVID-related restrictions, urban residents highlight growing expectations on urban space and micropolitics. Building on established debates on urban and infrastructure development and appropriation, the author makes an empirically grounded claim beyond utopian descriptions of circulating techno-centred visions and deterministic views of urban innovation. In concluding, he offers reflections about what the entanglement of bodies, infrastructures, and platforms through everyday life and survival mean for planning and theorizing the 'post-COVID' city and city of future.

In Chapter 10, Jonathan Cinnamon examines how scale has been mobilized as an analytical framework in urban data research, and what happens when the politics of urban data meets the politics of scale. A materialist framing provides a way of probing seemingly dissimilar concepts – quantitative data and geographic scale – as actors each with the potential capacity to enact political goals. Drawing on ongoing research in Cape Town on grassroots activism around informal settlements, Cinnamon concentrates on a particular moment in South African cities when 'data' emerged as a powerful discursive and material object within civil society organizations and social movements working to challenge injustice. While South African social movements have traditionally deployed scalar tactics, including scale jumping and multiscalar conflict, to open up new political terrains, he shows how new data-driven tactics of auditing and counting took priority in the fight against spatial injustice during the 'data turn' of the 2010s. In revealing the limitations of data and a subsequent remobilization of scalar tactics in this context, this analysis links data at the grassroots level with the post-political urban condition, suggesting a need to consider what forms of politics data enables and what forms it forecloses.

In Chapter 11, Evan Blake, Nancy Odendaal, and Ola Söderström analyse the tactics of civil society organizations (CSOs) in three South African cities: Cape Town; Ekurhuleni, in the Gauteng City Region; and Buffalo City. Drawing on work on data politics, data activism, and postcolonial science and technology studies, they use the notion of 'conjugated knowledge positions' to open the reflection to data tactics as part of broader knowledge politics and envisage them as negotiated within a multi-actor game. Based on their case studies they show how CSO tactics are positioned along a spectrum between data power and knowledge power. Extending work on CSO urban data politics they conclude that South African CSOs have not rolled out and rolled back data-focused tactics as a consequence of moments of faith and disillusionment in the power of data, but rather mobilize data and other forms of knowledge according to local political contexts and interactional situations.

In Chapter 12, 'Epilogue: Data, Crisis, and Learning', Orit Halpern, engaging the terms Anthropocene, technosphere, and smartness, argues that thinking big data and crisis together opens an avenue to reimagining new ideas about human – and more than human (including technology) – agency and subjectivity. Extending from this observation, this concluding essay then turns to reflexively think with the authors in the book, in order to ask how the careful examinations of big data might challenge contemporary assumptions of technical determinism and reconfigure our understanding of the future or urban life(s). Arguably, the careful study of the materialities, practices, and discourses of big data disrupts technically determinist imaginaries that propagate inequity and violence in the name of avoiding a future always imagined as catastrophic.

In line with the general perspective of the book, Halpern's epilogue thus points to ways to think and act in the age of data power with and beyond narratives of surveillance capitalism and creative data agencies to navigate urban digital futures.

References

Agrawal, A. and Kumar, V. (2020) *Numbers in India's Periphery: The Political Economy of Government Statistics*. Cambridge: Cambridge University Press.

Amoore, L.A. (2011) 'Data derivatives: On the emergence of a security risk calculus for our times', *Theory, Culture & Society*, 28(6): 24–43.

Arora, P. (2016) 'Bottom of the data pyramid: Big data and the Global South', *International Journal of Communication*, 10: 1681–99.

Attoh, K., Wells, K. and Cullen, D. (2019) '"We're building their data": Labor, alienation, and idiocy in the smart city', *Environment and Planning D: Society and Space*, 37(6): 1007–24.

Barns, S. (2018) 'Smart cities and urban data platforms: Designing interfaces for smart governance', *City, Culture and Society*, 12: 5–12. https://doi.org/10.1016/j.ccs.2017.09.006

Barns, S. (2020) *Platform Urbanism: Negotiating Platform Ecosystems in Connected Cities*. London: Palgrave Macmillan.

Beraldo, D. and Milan, S. (2019) 'From data politics to the contentious politics of data', *Big Data & Society*, 6(2). https://doi.org/10.1177/2053951719885967

Bigo, D., Isin, E. and Ruppert, E. (2019) *Data Politics: Worlds, Subjects, Rights*, London: Routledge.

Biruk, C. (2018) *Cooking Data: Culture and Politics in an African Research World*, Durham, NC: Duke University Press.

Boltanski, L. and Thévenot, L. (2006) *On Justification: Economies of Worth*. Princeton, NJ: Princeton University Press.

Bowker, G.C., Baker, K., Millerand, F. and Ribes, D. (2010) 'Toward information infrastructure studies: Ways of knowing in a networked environment', in J. Hunsinger, L. Klastrup and M. Allen (eds) *International Handbook of Internet Research*, Dordrecht: Springer, pp 97–117.

Burgess, J. (2022) 'Everyday data cultures: beyond Big Critique and the technological sublime', *AI & Society: Journal of Knowledge, Culture and Communication*, 38: 1243–44.

Burgess, J., Albury, K., McCosker, A. and Wilken, R. (2022) *Everyday Data Cultures*, Oxford: Wiley.

Casilli, A.A. (2017) 'Digital labor studies go global: Toward a digital decolonial turn', *International Journal of Communication*, 11: 3934–54.

Castells, M. (2010) *The Rise of the Network Society, The Information Age: Economy, Society and Culture, vol. 1* (2nd edn), Oxford: Blackwell.

Cinnamon, J. (2020) 'Attack the data: Agency, power, and technopolitics in South African data activism', *Annals of the American Association of Geographers*, 110(3): 623–39.

Couldry, N. and Hepp, A. (2016) *The Mediated Construction of Reality*, Cambridge: Polity.

Couldry, N. and Powell, A. (2014) 'Big data from the bottom up', *Big Data & Society*, 1(2). https://doi.org/10.1177/2053951714539277

Datta, A. (2015) 'New urban utopias of postcolonial India: 'Entrepreneurial urbanization' in Dholera smart city, Gujarat', *Dialogues in Human Geography*, 5(1): 3–22.

Datta, A. (2019) 'Postcolonial urban futures: Imagining and governing India's smart urban age', *Environment and Planning D: Society and Space*, 37(3): 393–410.

Datta, A. (2023) 'The digitalising state: Governing digitalisation-as-urbanisation in the Global South', *Progress in Human Geography*, 47(1): 141–59.

de Certeau, M. (1984) *The Practice of Everyday Life*, translated by S. Rendall, Los Angeles: The University of California Press.

Degen, M.M. and Rose, G. (2022) *The New Urban Aesthetic: Digital Experiences of Urban Change*, London: Bloomsbury.

Desrosières, A. (1998) *The Politics of Large Numbers: A History of Statistical Reasoning*, translated by C. Naish, Cambridge, MA: Harvard University Press.

D'Ignazio, C. and Klein, L.F. (2020). *Data Feminism*, Cambridge, MA: MIT Press.

Furlong, K. (2020) 'Geographies of infrastructure II: Concrete, cloud and layered (in)visibilities', *Progress in Human Geography*, 45(1): 190–8. https://doi.org/10.1177/0309132520923098

Gabrys, J. (2014) 'Programming environments: Environmentality and citizen sensing in the smart city', *Environment and Planning D: Society and Space*, 32(1): 30–48.

Graham, S. and Marvin, S. (1996) *Telecommunications and the City: Electronic Spaces, Urban Places*, London: Routledge.

Graham, S. and Marvin, S. (2001) *Splintering Urbanism: Networked Infrastructures, Technological Mobilities and the Urban Condition*, London: Routledge.

Graham, S.D.N. (2005) 'Software-sorted geographies', *Progress in Human Geography*, 29(5): 562–80.

Guma, P.K. (2020) 'Incompleteness of urban infrastructures in transition: Scenarios from the mobile age in Nairobi', *Social Studies of Science*, 50(5): 728–50. https://doi.org/10.1177/0306312720927088

Guma, P.K. and Monstadt, J. (2021) 'Smart city making? The spread of ICT-driven plans and infrastructures in Nairobi', *Urban Geography*, 42(3): 360–81.

Halpern, O. and Mitchell, R. (2023) *The Smartness Mandate*. Cambridge, MA: MIT Press.

Hollands, R.G. (2008) 'Will the real smart city please stand up? Intelligent, progressive or entrepreneurial?', *City: Analysis of Urban Change, Theory, Action*, 12(3): 303–20.

Houston, L., Gabrys, J. and Pritchard, H. (2019) 'Breakdown in the smart city: Exploring workarounds with urban-sensing practices and technologies', *Science, Technology, & Human Values*, 44(5): 843–70.

Hui, Y. (2016) *On the Existence of Digital Objects*, Minneapolis, MN: University of Minnesota Press.

Isin, E. and Ruppert, E. (2020) 'The birth of sensory power: How a pandemic made it visible?', *Big Data & Society*, 7(2).

Kennedy, H., Poell, T. and van Dijck, J. (2015) 'Data and agency', *Big Data & Society*, 2(2). https://doi.org/10.1177/2053951715621569

Kitchin, R. (2014) *The Data Revolution: Big Data, Open Data, Data Infrastructures and Their Consequences*, London: Sage.

Koselleck, R. and Richter, M.W. (2006) Crisis. *Journal of the History of Ideas*, 67(2): 357–400.

Lefebvre, H. (1991) *The Production of Space*, Oxford: Blackwell.

Leszczynski, A. (2019) 'Glitchy vignettes of platform urbanism', *Environment and Planning D: Society and Space*, 38(2): 189–208.

Leszczynski, A. and Elwood, S. (2022) 'Glitch epistemologies for computational cities', *Dialogues in Human Geography*, 12(3): 361–78.

Lupton, D. (2018) 'How do data come to matter? Living and becoming with personal data', *Big Data & Society*, 5(2). https://doi.org/10.1177/2053951718786314

Malaby, T.M. (2009) *Making Virtual Worlds: Linden Lab and Second Life*, Ithaca, NY: Cornell University Press.

Milan, S. (2013) *Social Movements and Their Technologies: Wiring Social Change*, New York: Palgrave Macmillan.

Milan, S. and Treré, E. (2019) 'Big data from the South(s): Beyond data universalism', *Television & New Media*, 20(4): 319–35.

Milan, S. and van der Velden, L. (2016) 'The alternative epistemologies of data activism', *Digital Culture & Society*, 2(2): 57–74.

Noble, S.U. (2018) *Algorithms of Oppression: How Search Engines Reinforce Racism*, New York: New York University Press.

Pollio, A. (2019) 'Forefronts of the sharing economy: Uber in Cape Town', *International Journal of Urban and Regional Research*, 43(4): 760–75.

Power, M. (2022) 'Theorizing the economy of traces: From audit society to surveillance capitalism', *Organization Theory*, 3(3). https://doi.org/10.1177/26317877211052296

Pybus, J., Coté, M. and Blanke, T. (2015) 'Hacking the social life of Big Data', *Big Data & Society*, 2(2). https://doi.org/10.1177/2053951715616649

Ruppert, E., Isin, E. and Bigo, D. (2017) 'Data politics', *Big Data & Society*, 4(2). https://doi.org/10.1177/2053951717717749

Simone, A. and Rao, V. (2021) 'Counting the Uncountable: Revisiting Urban Majorities', *Public Culture*, 33(2): 151–60.

Söderström, O., Paasche, T. and Klauser, F. (2014) 'Smart cities as corporate storytelling', *City: Analysis of Urban Change*, 18(3): 307–20.

Srnicek, N. (2016) *Platform Capitalism*, Cambridge: Polity.

Stephansen, H.C. and Treré, E. (2019) *Citizen Media and Practice: Currents, Connections, Challenges*. London: Routledge.

Strengers, Y. and Kennedy, J. (2020) *The Smart Wife: Why Siri, Alexa, and Other Smart Home Devices Need a Feminist Reboot*, Cambridge, MA: MIT Press.

Thatcher, J., O'Sullivan, D. and Mahmoudi, D. (2016) 'Data colonialism through accumulation by dispossession: New metaphors for daily data', *Environment and Planning D: Society and Space*, 34(6): 990–1006.

van Dijck, J. (2014) 'Datafication, dataism and dataveillance: Big Data between scientific paradigm and ideology, *Surveillance & Society*, 12(2): 197–208.

Vanolo, A. (2014) 'Smartmentality: The smart city as disciplinary strategy', *Urban Studies*, 51(5): 883–98.

Wallerstein, I., Collins, R., Mann, M., Derluguian, G. and Calhoun, C. (2013) *Does Capitalism Have a Future?* Oxford: Oxford University Press.

Zuboff, S. (2019) *The Age of Surveillance Capitalism: The Fight for a Human Future at the New Frontier of Power*, London: Profile.

PART I

Frames

2

Urban Data Power: Capitalism, Governance, Ethics, and Justice

Rob Kitchin

Introduction

Data have long been an important means for understanding and managing cities. During the Enlightenment and the establishment of modernity, scientific advances and the growth of bureaucracy significantly expanded the role of data for monitoring and regulating populations and their activities (Desrosières, 1998). States widened the systematic recording of data, such as registering personal information, conducting surveys and censuses, and tracking administrative services such as taxation, welfare, education, and health (Koopman, 2019). Data became a key source of evidence for social policy and the functioning of economies. The growth of double-entry bookkeeping and new accounting practices drove data practices within companies (Porter, 1995), later accompanied by business intelligence services (Gross and Solymossy, 2016), with data themselves becoming a tradable commodity (Sylla, 2002). In all these cases, the data produced and their associated infrastructures and practices were the product of data politics and were used to exercise data power. That is, data were produced and utilized to achieve particular aims and objectives for the interests of selected constituencies.

In the digital era, the importance of data as a resource and commodity has multiplied. This is particularly the case over the past two decades, given mass datafication and the rapid growth of big data, and their increasingly central role in the administration and operations of the state and business. Datafication is the process whereby more and more aspects of everyday life are captured as data, primarily through their digital mediation (van Dijck, 2014). Big data are produced continually and are exhaustive to a system; that is, the data are not sampled but are generated in real time for every individual, object, and transaction within a domain (for example, an automatic number plate recognition system tracks every single vehicle, not a sample of them) (Kitchin, 2022). Big data are essential elements of most smart city systems (for instance, integrated control rooms, coordinated emergency management systems, intelligent transport systems, smart energy grids, smart lighting and parking, sensor networks, building management systems) and urban platforms (such as Uber or Airbnb) (Kitchin, 2014). They are increasingly being used in performance management systems in order to monitor and direct city service delivery in a timely manner, and for city benchmarking and policy making (Kitchin et al, 2015). Financial big data and algorithmic systems are pivotal to the practices of fast and speculative urbanism, in which urban development is accelerated and intensified through the rapid circulation of data and capital (Datta, 2017).

Urban big data have become essential for how cities are planned and managed, how services are operated, and how business takes place within and between locales. Big data systems exert significant data power; that is, they possess the capacity to influence and transform social and economic relations and activities (Ruppert et al, 2017). In other words, they determine the outcomes of decision-making and action, with differential effects: working for the benefit of some (usually those that own or run systems) at the expense of others. Data power is used to maintain control or extract profit, or to socially sort people along the lines of race, ethnicity, gender, class, sexuality, disability, and other social markers (Browne, 2015; Eubanks, 2018). For example, data in administrative systems determine the services and benefits citizens receive, and how they are governed, based on their characteristics and activities (Kitchin, 2022). Data within predictive policing systems, or within housing investment applications, direct which areas and populations are targeted for attention (Jefferson, 2018; Safransky, 2020). Data within locative media and urban platforms shape the information and offers shared with users, and seek to influence and nudge their behaviour (Barns, 2020). These systems are saturated in data politics relating to the contested ways in which data are produced and used, whose interests they serve, and how data power is challenged and resisted. Such data politics is reflected in the varying points of view, agendas, rationalities, ideologies, and negotiations associated with data-driven systems and the work they perform.

This chapter is centrally concerned with the data power and data politics of urban big data systems. It argues that urban data power is principally (re)produced to deepen the interests of states and their ability to manage urban life, and companies and their capacity to create and capture new markets and accumulate profit. In other words, it is deeply imbricated into the workings and reproduction of political economies, its deployment justified as a necessary means to tackle various urban crises and sustain growth. Indeed, a set of persuasive discursive regimes have been constructed regarding the deployment of big data systems that promote and make their logic and application seem like common sense and the preferable way to try to solve urban problems (Kitchin, 2022). For example, the data power exerted through smart city technologies is justified as necessary to tackle three significant challenges: widespread changes in patterns of population, particularly rural to urban migration, and subsequent resource pressures; global climate change and the need to produce more resilient cities; and fiscal austerity and the desire to create leaner governments and attract mobile capital (White, 2016). Smart city technologies, it is argued, will enhance productivity, competitiveness, efficiency, effectiveness, utility, value, sustainability, resilience, safety, and security through the harnessing of computationally produced data power. The next section details how data power is being claimed and exerted through the logics and practices of data capitalism, particularly with respect to urban platforms. This is followed by a discussion of how data-driven systems are shifting the nature of governmentality and governance, enacting new, stronger forms of data power, as well as transferring some aspects of municipal government and service delivery to companies. The chapter then considers how data power is being resisted and reconfigured through an engagement with the ideas of data ethics, data justice, data sovereignty, and the practices of data activism.

Data capitalism and the city

The relationship between capitalism and urban development has long been theorized. As Brenner et al (2012, 3) contend, cities 'are sculpted and continually reorganized in order to enhance the profit-making capacities of capital' since they are 'major basing points for the production, circulation, and consumption of commodities,' as well as themselves being intensely commodified. Capitalism prioritizes exchange-value (generating profit) over use-value (the satisfaction of basic needs) and operates largely for the benefit of a relatively small group of elite actors who own and control the means of production (Harvey, 1985). The use of digital infrastructures, systems, and platforms, in the guise of producing a smart city, is the latest attempt by capitalism to leverage the city as an accumulation strategy, with companies seeking to capture and sweat, or disrupt and replace, public assets and services through technology solutions, support local economic development and attract foreign direct investment, drive real-estate investment, and foster a neoliberal, market-orientated approach to urban governance (Hollands, 2008; Shelton et al, 2015). A key element of this accumulation strategy is the data power enabled by the logics and practices of data capitalism.

Data capitalism is a form of capitalism wherein value and profit are driven in the main, or in large part, by extracting value from data, and data are themselves a form of capital and are key assets for speculative investment, not simply a commodity that can be converted into monetary value (Sadowski, 2019). The imperative for data capitalism is to generate, circulate, and monetize data. Mass datafication and the rollout of data-driven systems are a means of capturing and monetizing activities that have to date been weakly commodified and leveraging additional value from those already in the fold of capitalism. This is the prime reason that companies are supporters of the open data agenda: not to facilitate transparency and participation, but to gain free access to a resource that can be transformed into a product (Bates, 2012). To maximize profit, data capitalism seeks to obtain data for minimum cost and extract as much value as possible. In many cases, the data are generated without remuneration for labour, with the subjects and producers of data passively participating or knowingly creating data for free as an inherent feature of the system or platform (by being present and performing an activity, or by clicking, swiping, typing, uploading) (Sadowski, 2019). Communal resources, such as social communication or a public street, are enclosed through digital mediation, and personal activity and information datafied.

For some, this process of accumulation through data dispossession can be understood as forms of modern-day colonialism, in which the extraction of data, and through it the further colonization of daily life by capitalist interests, works in similar ways to historical, imperialist appropriation of territory and resources (Thatcher et al, 2016; Couldry and Mejias, 2019). Within data colonialism, data power is highly asymmetrical, with a system or platform owner controlling its operation, and challenging exploitative practices is difficult given their configuration and management (West, 2019). For example, on a locative media platform such as Foursquare there is a marked division between those who control the means of production and those who must submit to data extraction to gain access to service, the latter of whom are simultaneously a consumer (user), producer (labourer), product (data), and target (of value extraction, for example, to be sorted, judged, and nudged). While it might seem that some services are free for consumers to use, a price is being paid, dictated on the terms and services of companies.

Urban platforms are profoundly data-driven and derive their revenue from data monetization (usually by producing advertising revenue or selling data on to third parties), along with taking a fee for any goods or services sold via the platform and attracting venture capital. Data that are sold are often purchased by data brokers who consolidate multiple streams of data, repackage them into new products, and offer data services, such as microtargeted advertising, demographic profiling of individuals and places, assessing creditworthiness and risk, and business and bespoke data analytics (Roderick, 2014). These products and services can have a profound effect on cities by shaping decision-making and investments, in turn reinforcing and deepening social and spatial divides. This is particularly evident with respect to housing and the use of various forms of big data in making decisions relating to credit, tenancy, speculation, evictions, (dis)investment, and transfer of use (such as to short-term lets) (Safransky, 2020; McElroy and Vergerio, 2022).

This social and spatial sorting results in those that are already marginalized in society experiencing a double form of data colonialism (Mann and Daly, 2019). As well as experiencing new forms of data power, data colonialism amplifies historical forms of colonization and practices of social and economic exclusion (Ricaurte, 2019). This is particularly evident with respect to race, where people of colour are subjected to new algorithmic forms of violence, which build on and extend traditional forms of structural violence (Benjamin, 2019). For example, predictive policing seeks to anticipate and prevent future crime by analysing a range of data, such as the location and perpetrators of recently committed crimes, along with a range of longitudinal data relating to crime patterns and local intelligence, to guide patrol routes and target potential suspects (Shapiro, 2020). The algorithms used have been trained using historical records of crime, yet these data contain systemic bias given that black people are more likely to have been stopped and searched, arrested, and incarcerated (Brayne, 2017). Older forms of bias and violence are encoded into new forms of structural violence, further targeting black people, recreating a self-fulfilling cycle, and perpetuating institutional racism (Jefferson, 2018; Moses and Chan, 2018). Smart city technologies produce what Benjamin (2019) terms a 'new Jim Code', an algorithmic version of the Jim Crow laws that enforced segregation. Rather than tackling crises of urban poverty, discrimination, and segregation, they help deepen them.

Service-orientated, data-driven smart city technologies typically generate revenue through service contracts with state bodies, and creating and selling derived data products. Along with accessing open data, these technologies enable capital to colonize state data, enclosing them within their data infrastructures, where they are transformed and value added to produce new services, the primary market for which is often the same state bodies from which they were extracted (Bates, 2012). At the same time, the delivery of public services becomes ever more orientated around the production and consumption of data, and the role of data intermediaries becomes normalized. The most recent corporate innovation to enact data capitalism is for companies to try to capture the role of the state, moving beyond supplying services to, or acting on behalf of, the state to become state-like and sovereign, owning and governing settlements (Sadowski, 2022). In effect, the state is transformed into a privately owned state-as-a-platform in which a company constructs and controls all aspects of a locale including territory, buildings, infrastructure, service delivery, and governance (Sadowski, 2022). These ambitions do not relate solely to utopian, separate, autonomous enclaves, campuses or company towns, but ordinary neighbourhoods in cities.

The most documented attempt to create such a state-as-a-platform neighbourhood is Quayside in Toronto, a waterfront development that was to be delivered by Sidewalk Labs, a subsidiary of Alphabet, Google's parent company (Hodson and McMeekin, 2021; Sadowski, 2022). Announced in October 2017 and abandoned in May 2020, it aimed to create 3.3 million square feet of residential, office, and commercial space on a site of 12 acres, with ambitions to scale to a further 800 acres of adjacent land (Moore, 2019). It promised to be a neighbourhood built from the internet up, using a suite of smart city technologies to run a data-driven city. Significantly, Sidewalk Labs proposed to manage service delivery, which would all be private (for example, charter rather than public schools), take on governance functions, shape local and city policy, self-regulate their endeavours, and levy taxes (Mann et al, 2020; Tenney et al, 2020; Hodson and McMeekin, 2021). Similarly, some of the fast urbanization and smart developments in Africa seek administrative autonomy and an 'extra-territorial status that enables property owners to assume the bureaucratic responsibilities and regulatory functions once reserved for exclusive control by municipal authorities' (Herbert and Murray, 2015, 475). In other words, the neighbourhood developments are privately owned and administered, with little to no state involvement in local services and infrastructure provision and governance, with data-driven systems being key to their operation. Clearly, such arrangements wield enormous data power that is largely out of reach of democratic politics.

Digital governance, governmentality, and the city

As the Toronto example highlights, data-driven digital systems, infrastructures, and platforms are having a profound effect on urban governance and governmentality. This is occurring in two interrelated ways. First, and dovetailing with the rise of data capitalism, is the deepening of the neoliberal agenda and the extension of the role of industry in working with or on behalf of states to deliver essential city services. Technology companies have been actively targeting municipal governments for business contending that their products and services can more effectively and efficiently solve urban issues and undertake the work traditionally performed by the state (Söderström et al, 2014; Sadowski and Bendor, 2019). A key element of their argument is that the public sector lacks the core skills, knowledge, and capacities to address pressing contemporary social issues and maintain critical services and infrastructures, which can only be provided by specialist enterprises, market-led innovation, and technically mediated solutions (Kitchin et al, 2017). As such, state-led universal provision needs to be replaced by services delivered through a competitive marketplace, enabled through deregulation, public-private partnerships, outsourcing, and privatization (Brenner et al,

2010). Further, the state is encouraged to support and promote this transition through policy, market subsidies, and investment.

This neoliberal agenda drives government into the embrace of data capitalism, creating new long-term markets for capital accumulation. Importantly, neoliberalism also recasts urban citizenship. Rather than citizenship being grounded in inalienable rights and the common good, it is orientated towards market principles, with citizens reframed as consumers who have freedom of choice, but also responsibilities and obligations to act as states and markets dictate (Brown, 2016). Individuals are expected to navigate and negotiate the provision of services based on personal, social, political, and economic capital, framed within constraints that seek to limit excessive discrimination and exploitation (Brown, 2016). Citizens in the smart city can thus freely select services as long as they can afford them and they comply with state laws and corporate terms and conditions (Cardullo and Kitchin, 2019).

Second, new data-driven algorithmic forms of urban governance are being introduced. On the one hand, these systems are being used to make municipalities more business-like in their operation, utilizing new data streams to implement performance management systems designed to monitor workers and service delivery, and control and regulate infrastructure, in order to improve the efficiency and productivity of government. In effect, government is adopting the logics and practices of business intelligence to guide organizational and operational concerns, utilizing instrumental techniques such as tracking indicators, dashboards, and benchmarking (Kitchin et al, 2015). On the other hand, smart systems are being used to manage and regulate populations in more technocratic, instrumental, and automated ways, and often in real time. The digital mediation of services, utilities, policing, and security using big data systems is enabling five significant interrelated transitions in how society is governed.

First, smart city technologies significantly increase the scale and scope of surveillance regimes within public and private space. The transition from analogue to digital, and from visual to multi-sensor capture enables a variety of data to be monitored in real time using new, more sophisticated means of identifying, monitoring, storing, and acting on data streams, including facial recognition technology. A good example of this transition is with respect to policing, with forces in the US installing new command-and-control centres which employ extensive multi-instrumented surveillance (such as high definition CCTV, shot-spotter sensors, drone cameras, bodycams, online community reporting, as well as scanning communications and social media) to influence social behaviour and direct on-the-ground policing (Brayne, 2017; Wiig, 2018). Second, digital technologies and systems increasingly capture users within their rule-set and operations. These operations dictate pathways and actions, with failure to comply blocking progress. For example, an online welfare portal only permits certain ways of navigating and responding to complete a process. The entire interaction can be continuously recorded and is reactive to an individual's behaviour, but outside their control (Cohen, 2013). Third, digital systems permit the algorithmic processing and analysis of data; they are able to sort, sift, analyse, and act on streams of data in a systematic, consistent manner. Proponents argue that this algorithmic approach produces an objective, neutral assessment based on the data only, removing human bias from decision-making. Fourth, due to their computational competencies, the systems can operate in automated, autonomous, and automatic ways, enabling data to be processed and acted upon in real time (Kitchin and Dodge, 2011). This greatly increases the extent of monitoring and control as the systems can continuously perform governance functions. Lastly, streams of big data and advanced data analytics allow predictive profiling and anticipatory forms of governance across a number of domains; that is, to anticipate what is likely to happen under different conditions and for different populations and to act in a pre-emptive manner (Shapiro, 2020). Predictive policing enacts anticipatory governance, seeking to proactively prevent crime from taking place, and its logics are increasingly being applied to welfare assessments, security screening, and emergency management (Eubanks, 2018).

These five features of digitally mediated governance are reshaping governmentality; that is, the logics, rationalities, and techniques that render societies governable and enable governance, as well as extending the extent to which individual behaviour is guided and determined by companies and their technologies. Until relatively recently, the dominant mode of governmentality was disciplinary in nature (Foucault, 1991) in which technologies monitor individual behaviour from an external vantage point, with the possibility of being caught transgressing social expectations and laws leading to a self-regulation of action. Despite the procedures and technologies put in place, monitoring was periodic and somewhat haphazard. The increase in digital surveillance and the advent of big data has widened, deepened, and intensified the data gaze (Beer, 2019) and works to extend self-disciplining and associated disciplining measures (Kitchin and Dodge, 2011). This is being complemented with a control mode of governmentality in which an individual is subject to constant monitoring and modulation of behaviour, as the means by which a task is completed is also the means of governance (Deleuze, 1992). Rather than behaviour being shaped by fear of surveillance and sanction, in control systems individuals are corralled and compelled to act in certain ways, their behaviour explicitly or implicitly steered or nudged (Davies, 2015). That is, they are not self-disciplining their behaviour in relation to an external gaze, but their behaviour is actively reshaped through its digital mediation. For example, the work of checkout operatives in supermarkets is no longer disciplined through the gaze of the supervisor or CCTV monitoring work rate; now, the mode of work – the scanning of items – becomes the mechanism of capturing and regulating behaviour, continually monitoring performance and informing the worker to speed up if the scan rate is too slow (Kitchin and Dodge, 2011). As Davies (2015) notes, smart city developments and technologies are designed to capture, modulate, and nudge behaviour. His example is Hudson Yards in New York, a development saturated in sensors and embedded computation designed to continually monitor and modulate behaviour of residents and workers.

The introduction and operation of systems designed to reconfigure governance is thoroughly infused with data politics and data power, given what is at stake with respect to governmentality, democracy, and ethics. As is evidenced in places such as Hong Kong, where smart city technologies have been an important element in the new security apparatus designed to quell the democracy movement, systems that facilitate capture and control, automation, and prediction have profound social and political impacts (Lee and Chan, 2018). The use in Europe and North America of mass surveillance, the militarization of policing, the erosion of privacy and human rights, and the continual drift of control creep (in which systems designed for one purpose are enrolled into another) highlights that concerns regarding the path of data power is not limited to authoritarian regimes (Graham, 2011; Kitchin, 2021). Indeed, there are a whole series of ethics concerns relating to unfair and discriminatory treatment enacted within smart city systems (Kitchin, 2016). Consequently, while companies and states dominate the discursive landscape, and largely set the parameters for prevalent data regimes, data power is being met with resistance and counter-narratives and actions by other stakeholders designed to transform how digital devices, systems, infrastructures, and platforms work and produce alternative urban data futures.

Data justice and the city

Data capitalism and data-driven forms of governance, and the associated shift in governmentality, clearly raise a number of concerns relating to uneven and unequal distribution and consequences of data power, and how such power often deepens rather than addresses urban crises. The lives of individuals and communities are impacted in ways that suit the desires of capital and state power, with data power reinforcing and reproducing iniquitous structural relations. Data power is not, however, simply accepted on its terms, but is countered by resistance that directly opposes its operations or subverts and transgresses its intent. This occurs in a number of ways that can be loosely grouped into data ethics and data justice approaches.

Data ethics consists, on the one hand, of normative thinking concerning data-driven technologies and their practices, and, on the other, applied ethics that seeks to translate normative ideas into practice action. Normative ethics generally consists of applying ideas related to what can be considered right or wrong to data-driven systems (Kitchin, 2022); for example, advocating the design and operation of systems that conform to ethical positions such as deontology, consequentialism, virtue ethics, and ethics of care, each of which prioritizes a different value of assessment: following agreed rules, consequence of outcomes, intent to do the right thing, and treating others as one would want to be treated (Vaughan, 2014). Advancing normative ethics in relation to data-driven technologies aims to shift the ethos, principles, and values underpinning their development. In applied terms, normative arguments are mobilized in counter-narratives to free-market and neoliberal ideologies of data capitalism and laissez-faire governance, usually employing ideas centred on transparency, accountability, fairness, access, equity, rights, and citizenship. These notions are translated into more concrete interventions such as policy, regulations, law, and governance and management arrangements, designed to put in place checks and balances to the excesses of capital and institutional power (Kitchin, 2022).

A variety of actors, such as community groups, activist networks, civil liberties NGOs, and progressive political parties, working at different scales from the local to global, are actively involved in formulating and enacting data ethics in order to limit and redistribute data power. For example, at the international scale, NGOs such as the Electronic Frontier Foundation, Privacy International, Amnesty International, and European Digital Rights campaign for policy and legislative interventions related to privacy and data protection. Their work has led to interventions such as GDPR (General Data Protection Regulation) in Europe that obligates data controllers and processors to treat data in defined ways and provides citizens with rights with respect to data related to them (Voigt and von dem Bussche, 2017). At the same time, companies and states have embraced the notion of data ethics as way of setting up acceptable bordering principles around how data should be treated and used, and reassuring the public that their concerns are being taken seriously. This often results in advocacy for market-led, self-regulation (Crain, 2018) or the establishment of ethics advisory networks or boards, such as the Cities Coalition for Digital Rights, Seattle Community Technology Advisory Board, Amsterdam Tada!, and Smart Dubai AI Ethics Board.

While data ethics are important for challenging and tempering data power, for some critics and stakeholder actors they do not go far enough. Data ethics, it is argued, is too narrow in conception, locates the sources of concern in individuals and technical systems rather than social structures, pursues instrumental and procedural solutions rather than systemic change, and is too easily co-opted by those whose practices they seek to transform

(Dencik et al, 2016; D'Ignazio and Klein, 2020). As such, rather than fundamentally challenging and reconfiguring data power, it is contended that data ethics merely curbs particular practices, rather than addressing the root, structural conditions that enable discriminatory and exploitative data work, and thus continue to serve the vested interests of companies and states (D'Ignazio and Klein, 2020). While data ethics provides some protections, it does little to roll back, or offer a genuine alternative to, the operations of data capitalism and state dataveillance. Data policy, regulations, laws, and governance models continue to enable data capitalism to monetize data and accumulate profit through data colonialization, and facilitates states to enact data-driven governance in ways that reproduce uneven and unequal social relations. As D'Ignazio and Klein, (2020, 60) argue, a compliance regime set on the terms of companies and states will not address the ways in which discriminatory and exploitative data power pervades data-driven systems; data ethics are merely 'technological Band-Aid[s] for [...] much larger problem[s]'. These Band-Aids tackle symptoms not root causes, and they provide a captivating diversion from addressing those root issues (Powles and Nissenbaum, 2018).

Instead, D'Ignazio and Klein (2020) call for data justice rooted in a different set of concepts (justice, oppression, equity, co-liberation, reflexivity, contextual integrity, *in addition to* ethics, bias, fairness, accountability, transparency, understanding algorithms), wherein data power is recognized as a structural relation that cannot be ameliorated at the technical or individual level alone. They hold that the concepts of data ethics are valuable and useful, but in and of themselves they will not produce fair and just data regimes. Instead, a more radical shift in thought and praxis is required if a more equitable digital society is to be realized. Data justice applies the theories of social justice to data-driven systems and processes (Dencik et al, 2016), mapping out the logics, structural conditions, and operations of data power, charting data harms and their consequences, scoping alternative data futures and how they might be produced, and examining how groups are working to enact data activism and claim data sovereignty (the ability to control their data relating to them). Typically, the underlying moral philosophy of data justice draws on feminism (D'Ignazio and Klein, 2020), Marxism (Sadowski, 2019), and critical social theory more broadly (for example, Dencik et al, 2016; Taylor, 2017), with five forms of data justice identified (Cinnamon, 2017; Heeks and Renken, 2018; Robinson and Franklin, 2020): *instrumental*, concerned with the fair use of data and just outcomes; *procedural*, focusing on harms produced through data practices and processes, the 'biases and inequalities baked directly into data' (Cinnamon, 2017, 622); *distributional*, and the equitable distribution of data, associated resources, and data-driven outcomes; *recognition*, and the enactment of equal respect, rights, and treatment across all data subjects; *representation*, ensuring equal voice and

31

ability to shape and challenge data power across all data subjects. These components of data justice have been examined with respect to smart cities and urban data power through the work of Cinnamon (2020), focusing on the data politics of services in Johannesburg and Cape Town, and Robinson and Franklin (2020) and their analysis of urban sensor networks in Newcastle and Chicago.

Data activism and advocacy is a means to seek data justice: to challenge and transform data power into more equitable arrangements. In broad terms, it take two main forms (Milan and van der Velden, 2016). Reactive data activism aims to challenge, reconfigure, and dismantle asymmetric data power through political protest, legal cases, and advocacy and lobbying for policy change and regulation. At its most radical edge, it could involve widespread civil disobedience, vandalism, and hacking, as with the Umbrella protests in Hong Kong, which in part sought to block mass surveillance and data-driven security (Lee and Chan, 2018). Proactive data activism aims to use data (open data and self-generated) as a resource for political action and social change (Milan and van der Velden, 2016). Such work includes civic hacking, hackathons, and citizen science, all a means by which citizens produce their own data-driven solutions to social issues. In a number of cases, advocacy and activist organizations enact both forms of data justice. For example, the Detroit Digital Justice Coalition and Stop LAPD Spying Coalition, aided by national-level bodies such as the American Civil Liberties Union and Data for Black Lives, fight to change discriminatory practices such as social sorting, redlining, and data-enabled institutional racism, and also use data to campaign for social change (Currie et al, 2016; Petty et al, 2018). In so doing they aim to claim data sovereignty; that is, assert some level of authority and control over the data that relate to them and how those data are generated and used (Kukutai and Taylor, 2016). Data sovereignty has its roots in the claims of Indigenous peoples to the right to maintain, control, and protect their cultural heritage, traditional knowledge, and territories, and determine and govern how data related to these are produced, used, and shared (Mann and Daly, 2019). Such rights have long been denied, with data being extracted without consent within colonial relations for ends that rarely have been to their benefit (Kukutai and Taylor, 2016).

While the discussion so far has largely been a dualistic characterization of data ethics and data justice in opposition to data power, it is important to note that power, including data power, is never a simple binary of domination and oppression, imposition and opposition (Sharp et al, 2000), but is relational and entangled, often being fragmentary, uneven, inconsistent, and paradoxical. Individuals and institutions can simultaneously wield and be subject to various forms of data power. For example, a municipal worker might exercise data power in relation to a resident, but their own actions are subject to the monitoring of a line manager (which might be informed by feedback from citizens), and is sited within the governmentality of the institution, and local and national systems of oversight and benchmarking. Municipal workers both express and resist data power, with Kitchin et al (2017) noting the internal politics and contestation between units and staff within municipalities regarding smart city developments. This was also evident in the Toronto case, with an entangled, relational field of power struggles occurring within and between various bodies: corporations, municipal bodies, semi-state agencies, community groups, political parties, professional associations, university institutes, and others (Tenney et al, 2020; Hodson and McMeekin, 2021). Data power, then, is not imposed unilaterally, countered by data justice, but unfolds through complex relations of negotiation, persuasion, coercion, intimidation, alliances, betrayal, protest, advocacy, avoidance, subversion, and other tactics, between various constituents. These tactics play out spatially, with data–driven smart city initiatives 'subject to various territorializing and deterritorializing processes whereby local control is fixed, claimed, challenged, forfeited and privatized' (Duncan, 1996, 129).

While it is tempting to cast the Toronto case as a 'David' (community opposition) slayed 'Goliath' (Google Sidewalk Labs) tale in which data justice triumphed over data power, in reality it was a much more entangled, relational story in which various coalitions of actors sought differing outcomes, ranging from conditional support if changes were made to calls to end development. On the oppositional side, two coalitions included an independent lobby group, Toronto Open Smart Cities Forum founded by a university research centre, and a resident-led protest group, #BlockSidewalk, which sought to enact data justice. To counter their challenges, Sidewalk Labs appointed a digital strategy advisory panel, a data governance advisory working group, an advisory council of Canadian urban thinkers, a residents reference panel, and ran civic labs forums open to any member of the public (Vincent, 2019). A number of members of the Sidewalk Labs initiatives hoped to be able to shift the corporation's thinking and actions from the inside through their participation, but subsequently resigned over concerns with how the project was unfolding (O'Shea, 2018). This entangled field of relations stifled progress and led to Sidewalk Labs withdrawing from the Quayside development. However, it did so with the clear intention of trying again elsewhere, rather than folding or fundamentally shifting its smart city model. It is important then to be sensitive to, and unpack carefully, these relational and spatial operations of data power; in part, to detail the complexities of 'actually existing smart urbanism' (Shelton et al, 2015) and its data power, but also to provide insights into effective tactics for how data justice can be successfully achieved.

Conclusion

Urban data power is the product of political economies, mobilized to assert the entwined interests of states and companies, supposedly in order to tackle urban crises. Data power is central to the machinations of data capitalism, expressed through the asymmetric relations of data colonialism and the desires to accumulate through data dispossession and development and expansion of new data products, services, and markets. States leverage data power to more effectively and efficiently monitor and regulate populations, deepening regimes of surveillance and enabling a transition to control governmentality. Likewise, companies are using data-driven systems to govern worker performance in order to increase productivity. Data power, however, is not a unidirectional force, exerted as domination over weaker groups. Rather, data power is relational, contingent, contextual, and entangled in complex ways, and is variously scaled from the local to the global. Consequently, just as there are varieties of capitalism (Peck and Theodore, 2007), neoliberalism (Brenner et al, 2010), and smart urbanism (Caprotti and Cowley, 2019), there are varieties of data power associated with them.

Forms and expressions of data power vary in line with political economies and other axes of power such as nationalism. While many of the infrastructures, systems, and practices utilized are the same, how data power is mobilized, exerted, and its consequences, differ between democratic and authoritarian regimes. Mass state surveillance in China, and the deep interlinkages between state and corporate dataveillence, notably in its diverse social credit scoring apparatus (Liang et al, 2018), has a different character to the fractured state surveillance and its disconnect from corporate data regimes in Europe, where GDPR (and prior to that Fair Information Practice Principles) limits a state–industry data nexus (Kitchin, 2022). The ability to seek data ethics and data justice, and to practise data activism and claim data sovereignty, is also markedly different, with the Chinese state limiting and punishing opposition to its data regimes. This has been particularly evident in its handling of the democracy movement in Hong Kong and the installation of an extensive data-driven surveillance and security apparatus to quash dissent (Lee and Chan, 2018; Liao, 2020). Of course, data justice and activism are also opposed in the West by those that gain through data power, but there is more scope to fight for change without severe penalty.

Similarly, urban data power varies contextually, with the visions, objectives, and systems deployed varying across jurisdictions. In India, the 100 Smart Cities programme is part of a political, nationalist development agenda (Datta, 2018). In the UK, smart cities are part of a shift to a technocratic, neoliberal governance regime and demonstrator initiatives for exportable business opportunities (Caprotti and Cowley, 2019). In Germany, smart cities are about efficiency of urban governance and sustainable growth (Skou and Echsner-Rasmussen, 2015). In Japan, smart cities aim to address sustainability and create adaptive environments for an ageing population (Trencher and Karvonen, 2019). Within jurisdictions, smart urbanism varies based on the political regime, political administrative geography, state apparatus and governance structures, resources, and capacities of cities. For example, urban data power associated with the smart city initiatives of Barcelona shifted markedly with the change in government in 2015, with a neoliberal vision of a smart city replaced by a socialist view and the adoption of the principles of technological sovereignty (that systems have to reflect and prioritize the needs of citizens not corporations and states), open access data, software and infrastructure, and extensive citizen engagement in decision-making (Charnock et al, 2021). The fractured political administration of Metropolitan Boston, with its 101 autonomous towns and cities, limits data power at the metro scale, instead decentralizing it locally, where it can be variously expressed (Kitchin and Moore-Cherry, 2021). This is quite different to cities with a unitary metropolitan governance, such as New York, where data power is unified across the city-region.

Urban data power, then, needs to be examined and theorized within these terms: as a political-economic set of contingent, contextual forces and relations. Such work requires the carefully teasing out of its general characteristics and how it is mobilized, utilized, and contested in specific cases.

Acknowledgements

The research in this chapter was funded by a European Research Council Advanced Investigator Award, 'The Programmable City' (ERC-2012-AdG-323636). The chapter draws on a number of previously published works detailed in the references, but particularly the book, *The Data Revolution*, 2nd edition (2022).

References

Barns, S. (2020) *Platform Urbanism: Negotiating Platform Ecosystems in Connected Cities*, London: Palgrave Macmillan.

Bates, J. (2012) '"This is what modern deregulation looks like": Co-optation and contestation in the shaping of the UK's Open Government Data Initiative', *Journal of Community Informatics*, 8(2). https://doi.org/10.15353/joci.v8i2.3038

Beer, D. (2019) *The Data Gaze: Capitalism, Power and Perception*, London: Sage.

Benjamin, R. (2019) *Race After Technology: Abolitionist Tools for the New Jim Code*, Cambridge: Polity Books.

Brayne, S. (2017) 'Big data surveillance: The case of policing', *American Sociological Review*, 82(5): 977–1008.

Brenner, N., Peck, J. and Theodore, N. (2010) 'Variegated neoliberalization: Geographies, modalities, pathways', *Global Networks*, 10(2): 182–222.

Brenner, N., Marcuse, P. and Mayer, M. (2012). 'Cities for people, not for profit: An introduction', in N. Brenner, P. Marcuse and M. Mayer (eds) *Cities for People, Not For Profit: Critical Urban Theory and the Right to the City*, New York: Routledge, pp 1–10.

Brown, W. (2016) 'Sacrificial citizenship: Neoliberalism, human capital, and austerity politics', *Constellations*, 23(1): 3–14.

Browne, S. (2015) *Dark Matters: On the Surveillance of Blackness*, Durham, NC: Duke University Press.

Caprotti, F. and Cowley, R. (2019) 'Varieties of smart urbanism in the UK: Discursive logics, the state and local urban context', *Transactions of the Institute of British Geographers*, 44(3): 587–601.

Cardullo, P. and Kitchin, R. (2019) 'Smart urbanism and smart citizenship: The neoliberal logic of "citizen-focused" smart cities in Europe', *Environment and Planning C: Politics and Space*, 37(5): 813–30.

Charnock, G. March, H. and Ribera-Fumaz, R. (2021) 'From smart to rebel city? Worlding, provincialising and the Barcelona Model', *Urban Studies*, 58(3): 581–600.

Cinnamon, J. (2017) 'Social injustice in surveillance capitalism', *Surveillance & Society*, 15(5): 609–25.

Cinnamon, J. (2020) 'Attack the data: Agency, power, and technopolitics in South African data activism', *Annals of the American Association of Geographers*, 110(3): 623–39.

Cohen, J.E. (2013) 'What privacy is for', *Harvard Law Review*, 126(7): 1904–33.

Couldry, N. and Mejias, U.A. (2019) 'Data colonialism: Rethinking big data's relation to the contemporary subject', *Television & New Media*, 20(4): 336–49.

Crain, M. (2018) 'The limits of transparency: Data brokers and commodification', *New Media & Society*, 20(1): 88–104.

Currie, M., Paris, B.S., Pasquetto, I. and Pierre, J. (2016) 'The conundrum of police officer-involved homicides: Counter-data in Los Angeles County', *Big Data & Society*, 3(2). https://doi.org/10.1177/2053951716663566

Datta, A. (2017) 'Introduction: Fast cities in an urban age', in A. Datta and A. Shaban (eds) *Mega-Urbanization in the Global South: Fast Cities and New Urban Utopias of the Postcolonial State*, London: Routledge, pp 1–27.

Datta, A. (2018) 'The digital turn in postcolonial urbanism: Smart citizenship in the making of India's 100 smart cities', *Transactions of the Institute of British Geographers*, 43(3): 405–19.

Davies, W. (2015) 'The chronic social: Relations of control within and without neoliberalism,' *New Formations*, 2014(84–85): 40–57.

Deleuze, G. (1992) 'Postscript on the societies of control,' *October*, 59: 3–7.

Dencik, L., Hintz, A. and Cable, J. (2016) 'Towards data justice? The ambiguity of anti-surveillance resistance in political activism', *Big Data & Society*, 3(2). https://doi.org/10.1177/2053951716679678

Desrosières, A. (1998) *The Politics of Large Numbers: A History of Statistical Reasoning*, translated by C. Naish, Cambridge, MA: Harvard University Press.

D'Ignazio, C. and Klein, L.F. (2020) *Data Feminism*. Cambridge, MA: MIT Press.

Duncan, N. (1996) 'Renegotiating gender and sexuality in public and private spaces' in N. Duncan (ed) *BodySpace: Destabilizing Geographies of Gender and Sexuality*, London: Routledge, pp 127–45.

Eubanks, V. (2018) *Automating Inequality: How High-Tech Tools Profile, Police, and Punish the Poor*, New York: St Martin's Press.

Foucault, M. (1991) 'Governmentality', in G. Burchell, C. Gordon and P. Miller (eds) *TheFoucault Effect: Studies in Governmentality*, Chicago: University of Chicago Press, pp 87–104.

Graham, S. (2011) *Cities Under Siege: The New Military Urbanism*, London: Verso.

Gross, A. and Solymossy, E. (2016) 'Generations of business information, 1937–2012: Moving from data bits to intelligence', *Information & Culture*, 51(2): 226–48.

Harvey, D. (1985) *The Urbanization of Capital*, Oxford: Blackwell.

Heeks, R. and Renken, J. (2018) 'Data justice for development: What would it mean?' *Information Development*, 34(1): 90–102.

Herbert, C.W. and Murray, M.J. (2015) 'Building from scratch: New cities, privatized urbanism and the spatial restructuring of Johannesburg after Apartheid', *International Journal of Urban and Regional Research*, 39(3): 471–94.

Hodson, M. and McMeekin, A. (2021) 'Global technology companies and the politics of urban socio-technical imaginaries in the digital age: Processual proxies, Trojan horses and global beachheads', *Environment and Planning A: Economy and Space*, 53(6): 1391–411.

Hollands, R.G. (2008) 'Will the real smart city please stand up? Intelligent, progressive or entrepreneurial?', *City: Analysis of Urban Change, Theory, Action*, 12(3): 303–20.

Jefferson, B.J. (2018) 'Predictable policing: Predictive crime mapping and geographies of policing and race', *Annals of the American Association of Geographers*, 108(1): 1–16.

Kitchin, R. (2014) 'The real-time city? Big data and smart urbanism', *GeoJournal*, 79(1): 1–14.

Kitchin, R. (2016) 'The ethics of smart cities and urban science', *Philosophical Transactions of the Royal Society A*, 374(2083). https://doi.org/10.1098/rsta.2016.0115

Kitchin, R. (2021) *Data Lives: How Data Are Made and Shape Our Lives*, Bristol: Bristol University Press.

Kitchin, R. (2022) *The Data Revolution: A Critical Analysis of Big Data, Open Data and Data Infrastructures* (2nd edn), London: Sage.

Kitchin, R. and Dodge, M. (2011) *Code/Space: Software and Everyday Life*, Cambridge, MA: MIT Press.

Kitchin, R., Lauriault, T.P. and McArdle, G. (2015) 'Knowing and governing cities through urban indicators, city benchmarking and real-time dashboards', *Regional Studies, Regional Science*, 2(1): 6–28.

Kitchin, R., Coletta, C., Evans, L., Heaphy, L. and MacDonncha, D. (2017) 'Smart cities, epistemic communities, advocacy coalitions and the "last mile" problem', *it – Information Technology*, 59(6): 275–84.

Kitchin, R. and Moore-Cherry, N. (2021) 'Fragmented governance, the urban data ecosystem and smart city-regions: The case of Metropolitan Boston', *Regional Studies*, 55(12): 1913–23.

Koopman, C. (2019) *How We Became Our Data: A Genealogy of the Informational Person*, Chicago: University of Chicago Press.

Kukutai, T. and Taylor, J. (2016) 'Data sovereignty for indigenous peoples: Current practice and future needs', in T. Kukutai and J. Taylor (eds) *Indigenous Data Sovereignty: Toward An Agenda*, Canberra: Australian National University Press, pp 1–22.

Lee, F.L.F. and Chan, J.M. (2018) *Media and Protest Logics in the Digital Era: The Umbrella Movement in Hong Kong*, Oxford: Oxford University Press.

Liang, F., Das, V., Kostyuk, N. and Hussain, M.M. (2018) 'Constructing a data-driven society: China's Social Credit System as a state surveillance infrastructure', *Policy and Internet*, 10(4): 415–53.

Liao, R. (2020) 'The tech industry comes to grips with Hong Kong's national security law', *TechCrunch*, [online] 8 July, available from: https://techcrunch.com/2020/07/08/hong-kong-national-security-law-impact-on-tech/

Mann, M. and Daly, A. (2019) '(Big) data and the north-*in*-south: Australia's informational imperialism and digital colonialism', *Television & New Media*, 20(4): 379–95.

Mann, M., Mitchell, P., Foth, M. and Anastasiu, I. (2020) '#BlockSidewalk to Barcelona: Technological sovereignty and the social license to operate smart cities', *Journal of the Association for Information, Science and Technology*, 71(9): 1103–15.

McElroy, E. and Vergerio, M. (2022) 'Automating gentrification: Landlord technologies and housing justice organizing in New York City homes', *Environment and Planning D: Society and Space*, 40(4): 607–26. https://doi.org/10.1177/02637758221088868

Milan, S. and van der Velden, L. (2016) 'The alternative epistemologies of data activism', *Digital Culture & Society*, 2(2): 57–74.

Moore, M. (2019) 'Would you let Google run your city?' *Prospect*, 28 January 2019. https://www.prospectmagazine.co.uk/magazine/would-you-let-google-run-your-city

Moses, L.B. and Chan, J. (2018) 'Algorithmic prediction in policing: Assumptions, evaluation, and accountability', *Policing and Society*, 28(7): 806–22.

O'Shea, S. (2018) 'Ann Cavoukian, former Ontario privacy commissioner, resigns from Sidewalk Labs'. *Global News*, [online] 21 October, available from: https://globalnews.ca/news/4579265/ann-cavoukian-resigns-sidewalk-labs/

Peck, J. and Theodore, N. (2007) 'Variegated capitalism', *Progress in Human Geography*, 31(6): 731–72.

Petty, T., Saba, M., Lewis, T., Gangadharan, S.P. and Eubanks, V. (2018) *Our Data Bodies: Reclaiming Our Data*, Our Data Bodies project, [online], available from: https://www.odbproject.org/wp-content/uploads/2016/12/ODB.InterimReport.FINAL_.7.16.2018.pdf

Porter, T.M. (1995) *Trust in Numbers: The Pursuit of Objectivity in Science and Public Life*, Princeton, NJ: Princeton University Press.

Powles, J. and Nissenbaum, H. (2018) 'The seductive diversion of "solving" bias in artificial intelligence', *OneZero*, Medium, [online] 7 December, available from: https://onezero.medium.com/the-seductive-diversion-of-solving-bias-in-artificial-intelligence-890df5e5ef53

Ricaurte, P. (2019) 'Data epistemologies, the coloniality of power, and resistance', *Television & New Media*, 20(4): 350–65.

Robinson, C. and Franklin, R.S. (2020) 'The sensor desert quandary: What does it mean (not) to count in the smart city?' *Transactions of the Institute of British Geographers*, 46: 238–54.

Roderick, L. (2014) 'Discipline and power in the digital age: The case of the US consumer data broker industry', *Critical Sociology*, 40(5): 729–46.

Ruppert, E., Isin, E. and Bigo, D. (2017) 'Data politics', *Big Data & Society*, 4(2). https://doi.org/10.1177/2053951717717749

Sadowski, J. (2019) 'When data is capital: Datafication, accumulation, and extraction', *Big Data & Society*, 6(1). https://doi.org/10.1177/2053951718820549

Sadowski, J. (2022) '"State-as-a-platform": Sovereignty and capital in smart governance'. Paper presented at Beyond Smart Cities Today, Malmö, Sweden, 9–10 June 2022. https://www.iuresearch.se/videos/tEiAEv1FjKo/

Sadowski, J. and Bendor, R. (2019) 'Selling smartness: Corporate narratives and the smart city as a sociotechnical imaginary', *Science, Technology, & Human Values*, 44(3): 540–63.

Safransky, S. (2020) 'Geographies of algorithmic violence: Redlining the smart city', *International Journal of Urban and Regional Research*, 44: 200–18.

Shapiro, A. (2020) *Design, Control, Predict: Logistical Governance in the Smart City*. Minneapolis, MN: University of Minnesota Press.

Sharp, J.P., Routledge, P., Philo, C. and Paddison, R. (2000) 'Entanglements of power: Geographies of domination/resistance', in J.P. Sharp, P. Routledge, C. Philo and R. Paddison (eds) *Entanglements of Power: Geographies of Domination/Resistance*, London: Routledge, pp 1–42.

Shelton, T., Zook, M. and Wiig, A. (2015) 'The "actually existing smart city"', *Cambridge Journal of Regions, Economy and Society*, 8: 13–25.

Skou, M. and Echsner-Rasmussen, N. (2015) 'Smart cities around the world', *Geoforum Perspektiv*, 14(25): 61–7.

Söderström, O., Paasche, T. and Klauser, F. (2014) 'Smart cities as corporate storytelling', *City: Analysis of Urban Change*, 18(3): 307–20.

Sylla, R. (2002) 'An historical primer on the business of credit rating', in R.M. Levich, G. Majnoni and C.M. Reinhart (eds) *Ratings, Rating Agencies and the Global Financial System*, Boston, MA: Springer, pp 19–40.

Taylor, L. (2017) 'What is data justice? The case for connecting digital rights and freedoms globally', *Big Data & Society*, 4(2). https://doi.org/10.1177/2053951717736335

Tenney, M., Garnett, R. and Wylie, B. (2020) 'A theatre of machines: Automata circuses and digital bread in the smart city of Toronto', *The Canadian Geographer/Le Géographe canadien*, 64(3): 388–401.

Thatcher, J., O'Sullivan, D. and Mahmoudi, D. (2016) 'Data colonialism through accumulation by dispossession: New metaphors for daily data', *Environment and Planning D: Society and Space*, 34(6): 990–1006.

Trencher, G. and Karvonen, A. (2019) 'Innovating for an ageing society: Insights from two Japanese smart cities', in A. Karvonen, F. Cugurullo and F. Caprotti (eds) *Inside Smart Cities: Place, Politics and Urban Innovation*, London: Routledge, pp 258–74.

van Dijck, J. (2014) 'Datafication, dataism and dataveillance: Big Data between scientific paradigm and ideology, *Surveillance & Society*, 12(2): 197–208.

Vaughan, L. (2014) *Beginning Ethics: An Introduction to Moral Philosophy*, New York: W.W. Norton.

Vincent, D. (2019) 'Newly formed citizens group aims to block Sidewalk Labs project', *Toronto Star* [online] 25 February, available from: https://www.thestar.com/news/gta/2019/02/25/newly-formed-citizens-group-aims-to-block-sidewalk-labs-project.html

Voigt, P. and von dem Bussche, A. (2017) *The EU General Data Protection Regulation (GDPR): A Practical Guide*, Cham: Springer.

West, S.M. (2019) 'Data capitalism: Redefining the logics of surveillance and privacy', *Business & Society*, 58(1): 20–41.

Wiig, A. (2018) 'Secure the city, revitalize the zone: Smart urbanization in Camden, New Jersey', *Environment and Planning C: Politics and Space*, 36(3): 403–22.

White, J.M. (2016) 'Anticipatory logics of the smart city's global imaginary', *Urban Geography*, 37(4): 572–89.

3

Platforms as States: The Rise of Governance through Data Power

Petter Törnberg

Introduction

Recent years have seen the explosive growth of platforms such as Amazon, Alphabet, Airbnb, Facebook, and Uber – forming an ecosystem which is now central to contemporary capitalism while amassing unprecedented levels of money and influence (Langley and Leyshon, 2017; van Dijck et al, 2018). As any fundamental shift, platformization is born and shaped from crisis. The Great Depression birthed Fordist–Keynesianism, the 1970s crisis brought post-Fordism and neoliberalism (Harvey, 2007), and the first steps of the incipient rise of a digital form of accumulation was birthed in the 2008 financial crisis. Its dominance was cemented and made visible through the COVID-19 pandemic, and it appears now to be maturing through the subsequent financial and inflation crises. The large core corporations left following this process are emerging as a new form of 'company-states': firms with the capacity to control not only trade but also law, territory, and liberty – in other words, to regulate life. This role has not escaped the firms themselves, many of which view their governance as so central to their business model that they refer to their users as 'citizens'. Platformization thus signifies a transformation of urban governance and politics, as 'data is generative of new forms of power relations and politics' (Bigo et al, 2019, 4).

In this chapter, we will examine the impact of data's transformation of governance. We here view the growing powers of data to shape human life as lying at the core of platformization. On the basis of this perspective, the chapter will examine platformization in two parts.

First, we will examine platformization as a form of capitalist accumulation based on employing data power to privatize regulation. Platformization first emerged as *proprietary* markets owned and created by platform corporations – such as Airbnb or Uber (Langley and Leyshon, 2017). The proprietary market business model is based on using the control over markets to extract monopoly rents. This has been described as a continuation of neoliberalism's constant annexation of new fields by the market, reaching its logical endpoint in the market's annexation of *the market itself* (Barns, 2020). However, the logic of platformization has since generalized to the use of data power to manipulate markets in order to extract profits through the concentration of political-economic power. Platformization can thus be understood as private actors employing digital technopolitical strategies to target vulnerabilities in local institutions, in the pursuit of control over market regulation. Platforms seek to claim regulatory control through data surveillance, while seceding from state control, thereby challenging the distinction between the economic power of corporations and the political sovereignty of states. The result is a gradual and variegated shift towards the private capture of governance, as capital supplants democratic institutions with private technological solutions.

Second, we will examine the nature of regulation as it is pursued through data power. Platform regulation implies a fundamental shift in the *way of seeing* those governed, bringing a shift in the regime of power. Scott (1998) famously characterized how the modernist state made the social world legible and amenable to state power through a top-down population-based epistemology, exerting power through hierarchical command-and-control that spread from the Fordist factory to shaping society, cities, and even a period of modernity. The platform mode of regulation implies a new way of seeing, as platforms see those governed through the novel epistemology of Big Data – cluster-based, bottom-up, and relational – and exerting control through the design of programmable social infrastructures. This signifies a fundamental shift in the regime of power, lying at the heart of the societal transformations emerging from digitalization and platformization.

Accumulation through data power

Digitalization first emerged as part of the macro-trends of capitalist reorganization that followed the Fordist crisis of the 1970s: financialization, globalization, and neoliberalization. Digital technology provides the infrastructure for the global financial system, as financial products are fundamentally predictive mathematical and computational entities. The growing sophistication of digital data and algorithms enable the financialization and annexation into capitalism of an expanding field of social behaviour (Sadowski, 2019) – 'liquifying' areas previously inaccessible to capital (Lohr, 2015; van Dijck, 2014).

But while digitalization was part and parcel of these macro-trends, it also brought with it challenges to existing capitalist institutions: as digital goods are not scarce but can be copied with near-zero marginal costs, they pose a much debated dilemma for accumulation. By bringing an end to the scarcity on which profits depend, some scholars even speculated that digital technology would bring the arrival of a postcapitalist utopia (Mason, 2016).

Capitalism's solution to the dilemma posed by digital technology was the *platform*: a natively digital organizational form, which allows the creation of artificial scarcity by using digital technology's capacity to centralize and control access to key resources. Platforms make use of the affordances of digital technology to curate programmable social infrastructures that enable buyers and sellers to meet; that is, to constitute a form of *proprietary market* (Langley and Leyshon, 2017). The platform business model can thus broadly be understood as leveraging digital technology to capture the market itself, and financialize its ownership and regulation.

The rise of proprietary markets can be seen as a fundamental transition in the structure of capitalist regulation: if Fordism was defined by national markets with national state regulation, and post-Fordism by transnational markets with national regulation, then digital capitalism is defined by *proprietary* markets – owned and regulated by transnational private companies through digital technology. Seen through this lens, the platform model constitutes the convergence of several long-running post-Fordist trends: neoliberalism's tendency to privatization and financialization of everything; the flexible formation of new financial conventions; use of digital code as means of shaping social institutions, and data as means of financializing them.

While platformization began with the proprietary markets of 'sharing economy' platforms such as Airbnb and Uber, it has been gradually evolving into a broader capitalist logic. At its core, the platform model is founded on leveraging data power as mechanisms for market dominance: platformization implies seeking to claim control over strategic chokepoints for accumulation, enabling firms to manipulate the market and extract rents from producers by controlling access. As Peck and Phillips (2020) argue, platforms can thus be understood as situated in the Braudelian zone of the 'antimarket', constituting a 'new machine with an old purpose: that of controlling markets from above and, in the process, generating significant concentrations of political-economic power' (p 75). While Fordism pursued profits through wealth creation and rationalization of production, and post-Fordism through financial markets and wealth relocation, digital capitalism thus generates profit through rentiership – based on the capacity to control access to key resources (Langley and Leyshon, 2017). The platform strategy thus hinges on using data power to make markets uncontestable by raising steep barriers to entry (Baumol, 1986), thus allowing the extraction of monopoly rents.

The monopolies of digital capitalism are thus fundamentally different from the steel and rail monopolies of the Fordist era: firms like Amazon are not even close to having a monopoly on retail – but are yet able to extract monopoly rents by drawing on data power (Peck and Phillips, 2020; Zuboff, 2019). Platforms use three forms of data power to achieve such market dominance. First, platforms use the strategic employment of *infrastructuralization* to produce lock-ins: platforms seek to provide basic functions that become entrenched, creating dependence on a privatized infrastructure (Larkin, 2013). As Rahman and Thelen (2019, 180) observe, 'the very idea of the "platform" reflects an aspiration to be the foundational infrastructure of a sector.' Second, the mediating position granted by ownership of infrastructures gives access to data flows, allowing platform companies to shape social pattern through global architectures of behavioural monitoring, analysis, prediction, and modification (Zuboff, 2019). The capacity to draw advantages from massive amounts of data, scalable at near-zero cost, results in feedback loops generating market concentration – what has been referred to as 'digital monopolies' or 'dataopolies'. Third, through the strategic employment of demand-side economies of scale – so-called 'network effects' (Rochet and Tirole, 2003): since the value of using a platform is a function of the number of market participants, incumbents are strongly favored (McAfee and Brynjolfsson, 2017). The result is a 'feedback loop that produces monopolies' (Parker et al, 2016, 6), leading to most mature platform markets being dominated by one or two giants (Peck and Phillips, 2020).

As platforms become truly valuable only if they can claim control over a key resource, competition plays out as winner-take-all turf wars that systematically favour capital and scale, in which dominant platforms leverage power in one sector to override competition in others (Cusumano et al, 2020). Unlike the monopolies of the Fordist era, the new form of monopoly power is not based on vertically integrated corporations and direct ownership, but on digital capacities for market control and manipulation (Peck and Phillips, 2020; Zuboff, 2019). The result is that corporations grow and expand according to a data-centric logic – continually spreading their roots to claim control of the infrastructure on which their rivals depend, and extend their data extraction into new areas – capturing and consolidating markets through what Srnicek (2017, 256) describes as a 'rhizomatic form of integration.' As a result of such horizontal expansion, platform firms spread and compete across a range of markets: Amazon (originally a bookstore), Google (originally a search engine), and Meta (originally a social networking website) are now engaged in turf wars to claim control over diverse market segments.

Platformization of regulation

As platforms are seeking to capture control over markets, the state is effectively part of their competition. As Kitchin (this book) notes, platform firms therefore 'try to capture the role of the state, moving beyond supplying services to, or acting on behalf of, the state to become state-like and sovereign, owning and governing settlements'. The well-documented regulatory and political consequences of platforms are thus part of platforms' competition with states, as platforms seek to exploit institutional weaknesses in order to break out of the control of the state.

Smart cities are among the clearest examples of such platformization of regulation in action. As Kitchin (this book) notes, they represent the strategy of capturing public services through technopolitical solutions, to form a market-orientated approach to urban governance – while generating revenue through service contracts with state bodies, and the extraction of citizen data. As the emerging literature on 'platform urbanism' highlights, the urban is central to the platform capitalist form of accumulation, with irreducible, co-generative dynamics between platforms and the city (Barns, 2020; van Doorn, 2019). Platforms are coming to 'alter the conditions through which society, space, and time, and thus spatiality, are produced' (Kitchin and Dodge, 2011, 13), using data as the new means to remake the city in capital's image (Couldry and Mejias, 2019).

The recent leak of internal files from Uber provides an example of the strategy that platforms pursue to achieve their political goals. As the files reveal, Uber seeks to exploit regulatory loopholes and mobilize political and legal power to avoid having its business regulated by states – strategically breaking laws, bypassing regulations, exploiting violence against drivers, and lobbying governments across the world (Davies et al, 2022).

The centrality of politics to platforms means that their impact and nature are highly contingent on the local institutional landscape, as they seek to target and exploit specific local regulatory and institutional conditions. For example, Thelen (2018) finds that the disruptive effects of Uber differ significantly across Germany, Sweden, and the United States, as the platform adapts to local forms of regulation and governance, seeking to identify and target loopholes and regulatory grey zones. Such differences highlight that platformization – as neoliberalization before it – is simultaneously novel, while contingent and path-dependent, and generating variegated outcomes.

While the impact of platformization varies markedly across territory, platforms tend to follow some common political strategies, enabled by the power of digital technology. We will here briefly outline key characteristics of the platform strategy vis-à-vis states, and how these strategies shape variegated pathways of regulatory transformation.

The platform model implies developing technological solutions to social problems – what Morozov (2013) calls 'technological solutionism' – as strategic means for colonizing the realm of political decision-making: supplanting public and political with private and technological. The platforms use technology to exploit regulatory grey areas – engaging in what Hecht (2000) calls 'technopolitics': the 'strategic practice of [...] using technology to constitute, embody, or enact political goals' (Hecht, 2000, 15). While platforms are better understood as regulatory than technology entrepreneurs (Pollman and Barry, 2016), technological innovation is thus central to their regulatory pursuits: while technological innovation always has had political consequences, those consequences have now increasingly become the chief purpose. Big Tech should thus in this sense be understood as representing the rise of a new form of technopolitical firms.

Platforms tend to seek rapid expansion, fueled by massive venture capital backing to undercut competition and quickly build a userbase (Langley and Leyshon, 2017). While the literature has understood this primarily as a means of outcompeting other platforms, it also serves as a strategy vis-à-vis the state, as quick expansion allows the platform to build political and legal power, hire lawyers and lobbyists, and mobilize its user base as a political force (Collier et al, 2018; van Doorn, 2019; Culpepper and Thelen, 2020). Having established a business in a regulatory grey area, the rapid expansion allows companies to present slow-moving lawmakers with a fait accompli, while mobilizing overwhelming political and legal power to fight attempts at after-the-fact regulation (Srnicek, 2016).

Platforms seek to spread their rhizomatic roots to claim control of the infrastructure on which states, regulatory agencies, and political elites depend. As platform corporations have thus emerged as the 'infrastructural core' (van Dijck et al, 2018, 12) of the global digital economy, they have also become embroiled in geopolitical conflict – emerging as 'key pawns in a mounting hegemonic strife' (Bassens and Hendrikse, 2022, 1), in particular between China and the US. As a result, states seek to support platformization as means of geopolitical influence (Peck and Phillips, 2020).

Platforms seek to avoid taxation and regulation by claiming to constitute a thin layer of intermediation which merely helps connect market actors. While platforms exert significant control through infrastructural design and data extraction, they often seek to maintain a narrative of neutrality in order to avoid regulatory responsibilities. For instance, labour platforms like Uber or MTurk draw on their algorithmic form of worker control (Cheney-Lippold, 2011) to claim that their workers are not employees, but 'independent contractors' who are therefore not fully subject to labour laws and welfare state protections (Ravenelle, 2019). This means that the platforms can devolve onto workers costs and risks such as varying demand, lost earnings, responsibility for bodily injury, and damage to tools and assets. Short-term rental platforms such as Airbnb likewise use similar narratives to claim that they are merely connecting guests to private home rental, thus bypassing the regulation of hotel accommodation and shifting responsibility for taxation and legal obligations to their 'hosts' (Törnberg, 2021). This is part of a broader strategy, in which platforms – from social media to gig work – use technological designs to target regulatory grey areas, algorithmic governmentality to shape the market to their interests, while drawing on a discourse of neutrality in order to shift legal responsibilities onto their users (van Dijck and Poell, 2013).

At the same time as platforms pass on regulation onto their users, they function as a legal and political front for these users – concealing their identities and mobilizing legal and political power to shield them from regulatory burden. Airbnb, for instance, has been shown to actively obfuscate host information to conceal their identity from governments and tax agencies, to mobilize significant lobbying efforts to fight stringent regulation, suing governments and tax agencies, and even to organize their users in 'social movements' to push their political interests (van Doorn, 2019). Platforms thus attempt to effectively *un*nest their proprietary markets from the larger public market of which they are part, making participants subject only to the taxation and governance imposed by the platforms themselves. Platforms, in other words, seek to operate on the same level as sovereign states – as managers and regulators of markets.

In summary, platformization can thus be understood as the rise of a form of accumulation based on rentiership, using data power to claim control over regulation. As Kitchin (this book) notes, 'the state is transformed into a privately owned state-as-a-platform in which a company constructs and controls all aspects of a locale including territory, buildings, infrastructure, service delivery, and governance'. Platforms are *technopolitical* actors, employing technology to constitute, embody, or enact political goals, seeking to employ digital power to capture and monopolize regulation. Through code and data, governance is depoliticized and put under private control, organized as proprietary algorithms which employ massive behaviour data to engineer social systems through infrastructural design – turning social issues into technical problems to be solved by private means.

Characterizing governance through data power: seeing like a platform

Scott (1998) famously characterized how the modernist state made the social world legible and amenable to state power through a top-down population-based epistemology, exerting power through hierarchical command-and-control. As Scott argued, any understanding of the world necessarily requires abstraction: a narrowing of vision to reduce the unwieldy complexity of reality into something manageable. Scott studied the way that the modern state made the social world legible, readable, and thus amenable to state power. The state needs maps to navigate and act – and to be useful, maps need to reduce and leave out.

In examining the emergence of the modern state, Scott found that the state's particular way of drawing its maps was at the core of a range of societal phenomena. When wielded by the state, a map becomes more than just a map: the world is reshaped and remade to fit the description that the map provides. A state registry which designates taxable property-holders does not merely *describe* a system of land tenure but *creates* such a system by giving its categories the force of law. Scott traced a long range of phenomena expressive of the modern state's particular way of rendering legible – from the standardization of weights and measures, the design of forests, the creation of permanent last names, cadastral surveys and population registers, to the grid design of cities. From these emerge a common pattern, in which the state fought against diversity, mobility, local traditions, and individuality, seeking to impose a world that matched the homogenizing rows and columns characterizing its ways of representing the world. These changes were attempts at making legible, taking complex and diverse local practices and slotting them into a standard grid whereby they could be centrally recorded and monitored.

As Bauman (2000) argued, the high modernist state's way of seeing emerged from the epistemic structure of the Fordist factory. Industrial capitalism was marked by the specialized division of labour, with specific characteristics: mass production based on standardization, rationalization, and the interchangeability of parts; mass production based on large groups of workers concentrated in factory settings, operating with functional specialization in administrative hierarchies and under strict managerial authority. The image of the Fordist factory shaped a modernity which was similarly obsessed with bulk and size, and impenetrable boundaries, with a preference for matching forms of planning and social organization – large factories and farms, huge dams, and grid cities.

Centrally, the modernist epistemology characterized a certain form of data, which matches the structure of the factory. It was the data of exact accounting, measurement, and statistics, printed sheets of IBM machines that governed every movement of the factory floor. The data of the 'average man', monitored through rows and columns of data, steered through top-down command-and-control. Drawing on such data, the image of the Fordist factory was transposed to society at large, institutionalized in schools, hospitals, family life, and personality – as figures like Robert McNamara brought to statecraft and warfare what they had learned from managing – through IBM machines, statistics, detailed control, and strict hierarchies – the factory.

The high modernist abstraction permeated the lived experience of its era, built around the image of the heavy machinery of the factory, with its precise structures and control. Society became factory-like, shaped by institutions that mirrored industrial organization: schools, hospitals, and even family life. Its science was that of average man and *homo economicus*, statistics, and systems theory, based on variable–variance analysis of representative samples of survey data. Its quantitative social sciences have dominated up until today, obsessed with measuring, classifying, and categorizing, finding regularities through means and variances, through assumptions of homogeneity and linearity.

The rising use of data power for governance which characterizes platformization is driving a shift in this epistemic foundation of modernity. As capital is using code to rewrite laws and employing the medium of digital technology to supplant the role of democratic institutions, platformization is bringing the rise of a new form of regulation. The platform mode of regulation comes with a particular *way of seeing* those governed, as the digital is coming to replace the Fordist factory as the chief 'epistemological building site' (Bauman 2000, 82) for contemporary modernity. As the high modernist way of seeing before it, data power is emerging as an incipient paradigm, reflected everywhere in society. As the logic of heavy machinery permeated first modernity, so does the logic of computer code and data coming to permeate the society of today. The rise of data power implies a shift in two deeply intertwined dimensions of power: the way of rendering legible, and the way of exerting power.

In terms of how platforms render legible, the shift consists of a move from traditional data to Big Data. This shift is not merely a question of new quantities of data or new tools – rather, in the words of Boyd and Crawford (2012), digital data are associated to 'a profound change at the levels of epistemology' (p 665). While survey data is constructed for processing through variable-based analysis, requiring pre-compartmentalized data designed to be palatable for a scientific perspective that sees the social world through a lens of averages and variances, Big Data tends to be structured by and for algorithmic processing, implying indexed data structures and traversable networks (Mackenzie, 2012; Marres, 2017). While traditional data slot reality into fixed categories, variables, and variances, concealing its interactional elements (Conte et al, 2012; Lazer et al, 2020), Big Data are relational, interactive, heterogeneous, interactional, and emergent (Törnberg and Uitermark, 2021a). The social ontology that digital technologies operationalize is not focused on the summing up of a population in fixed categories, but rather on the individuals and their dynamic connections and interactions (Uprichard, 2013; Castellani, 2014; Törnberg and Törnberg, 2018). Rather than focusing on populations – assumed to be the sum of their parts – Big Data sees the world through clusters and patterns, located within a larger data structure.

While traditional data was collected periodically, giving a snapshot of a defined population, Big Data is continuously gathered – and continuously fed algorithms that redefine clusters and patterns and seek to modulate their behaviour. Data power is fueled by a continuous flow of surveillance and control, from sensors that are seamlessly integrated into the urban fabric.

Big Data thus gives space for forms of diversity, mobility, and individuality that traditional data erased – tracing individuals through thousands of ever-shifting attributes. While traditional data sees order *from above*, digital data sees it *from below*: traditional data imposes grids and straight lines, while Big Data allows fractal structures and diversity. But the epistemic shift associated with Big Data representations does not imply that the world is more correctly represented: as new aspects are brought into focus, others become blurry (Andersson and Törnberg, 2018). Any way of rendering legible requires abstraction, erasing aspects of the phenomenon.

In terms of how platforms exert power, the shift consists of a move from top-down command-and-control to a form of control mobilized through the design of programmable social infrastructure. If 'the medium is the message', as McLuhan argued, then consequently the one who controls the medium controls the message. This is the foundation of platform power. Platforms operate by providing the social infrastructures that underlie actions, and thus exerting control by designing these infrastructures so as to generate certain outcomes – drawing on massive behaviour data to engineer social systems through infrastructural design. Yeung (2017) refers to this mode of control as 'hypernudging', as digital platforms engage in a rigorous process of designing the architectures to alter behaviour in predictable ways. Platforms shape their users through a mix between soft and hard discipline, combining gamification and scores with detailed tracking, algorithmic control, and at times threats of fines and expulsion – all A/B-tested and designed to efficaciously shape user behaviour.

To design infrastructures is to define the rules and goals of the social games that people are playing as they engage in the world. As Thi Nguyen (2020) argues, such games operate in the medium of agency: they have the power to determine not only the mode of interaction, but the goals and motivations of players – that is, to shape their very subjectivity. To control a social infrastructure is to gain some level of control over the goals and rules governing social life. This is not to suggest that data power vacates the role of individual agency – but rather to say that it situates and sets the context of agency. As Marx famously noted, we make our own history – but not under the circumstances of our choosing. Platform power implies control not over our choices, but over their circumstances, by defining the material life so central to conditioning social life.

While regulatory power targeted individuals, platform power thus operates on the interhuman and relational level, seeking to algorithmically modify the social rules that govern social behaviour. Platform power thus implies a relational approach to control, reshaping the connections and relations between people, leveraging social behaviour to generate social pressure for change. While an individual may, of course, choose not to play or to disregard the imposed rules of the game, this will, as in any game, inevitably imply losing in the eyes of those who are playing.

Twitter provides an example of this form of power in action (Nguyen, 2021; Törnberg and Uitermark, 2021b). When we engage in public conversation and discourse, we engage in a complex social activity in which each individual pursues their own goals – implicit, and often rich, subtle, and conflicting. Twitter's interface thus constitutes the most profitable answer to the question: what type of game is public discourse? Twitter not only defines how we interact and with whom, but centrally supplants this nuance and diversity with simple points-based scoring systems to measure our conversational success – retweets, likes, and followers. By defining measures of our success that are irresistible in their simplicity and clarity, Twitter re-engineers our communicative goals. The effects of this are not restricted to the confines of the platform itself, but as social media have become the chief engine of public discourse in our society, the aims and motivations seep out to redefine public discourse and even political life more broadly – in a process that Hepp (2020) refers to as 'deep mediatization'.

As Twitter applies this form of power to public conversations, so labour platforms like Uber are employing similar strategies for worker control. While purporting to provide a ride-share market, Uber sets the base rates its drivers charge, and limits the ability of drivers to accept or reject these offers – even creating 'phantom cabs' to give an illusion of greater supply to push down prices (Rosenblat and Stark, 2016). The Uber reputation system works as a normative apparatus, nudging both drivers and passengers toward a specific behaviour through scores, nudges, detailed tracking, algorithmic control, and threats of fines and expulsion – all A/B-tested and designed with precision to shape worker behaviour. At the same time, platforms shape subjectivities of workers by having them interact as competitors in a market rather than collaborators in a team, designing interfaces to prevent communication, and seeking to prevent emergence of a critical political subject needed for resisting the disembedding brought by the labour platform.

Törnberg and Uitermark (2020b) and Isin and Ruppert (2020) situate this novel digital form of control in Foucault's history of power, arguing that it signifies a move from regulatory power's top-down 'average man' data epistemology to a power shaped by the epistemic features of Big Data: cluster-based, relational, interactional, fluid, and ostensibly bottom-up. In the same way that Foucault (2008, 259) suggests that the modern disciplinary power was reshaped by the biopolitical power exerted by neoliberal rationalities, so is the biopolitical power of neoliberalism thus now being altered by the digital power made possible by digital technologies (Cheney-Lippold, 2011; Pfister and Yang, 2018). Platformization thus constitutes the rise of a new governing logic, coming to shift the fundamental market ideology, discipline, and rationality. As the neoliberal rationality came with an associated ideology and belief in the legitimacy of market rationality in regulating every aspect of human life, so does this complex control come with its associated ideology: what Malaby (2009) terms 'technoliberalism', defined by faith in the legitimacy of emergent effects – 'the emergent properties of complex interactions enjoy a certain degree of rightness just by virtue of being emergent' (Malaby, 2009, 56). That is, the trust in the invisible hand of the platform algorithm.

Isin and Ruppert (2020) use the notion 'sensory power' to refer to this novel regime of power – as it is characterized by data collected from sensors. As Isin and Ruppert (2020) note, these regimes of power should be understood as layered rather than consecutive: it is not that old forms of power fall into complete disuse and become replaced, but rather that new forms emerge alongside them, nestling and intertwining, varying in salience across periods and contexts. Törnberg and Uitermark (2020a) instead use the term 'complex control' to describe the emerging regime of power, suggesting that is should be understood through the epistemology of the digital. The epistemic nature of the digital can best be understood through the fundamental distinction between *complex* and *complicated* systems (Érdi, 2007; Andersson and Törnberg, 2018; Törnberg and Uitermark, 2021a). The epistemology described by Scott's characterization of the modern state was founded on complicatedness. Complicated systems are like sophisticated machineries: top-down, hierarchical and bureaucratic, each of their components designed to carry out an organized function that fits into a larger structure. Such systems can be made highly efficient and capable of executing large-scale tasks with extreme precision, but they are at the same time brittle: fragile to internal and external disturbances, and lacking in their capacity to adapt to shifting circumstances (Michod and Nedelcu, 2003). With the rise of digital power, we are seeing the shift to a complex regime of power. Complex systems tend to have less functionally differentiated components, and are instead organized through large sets of interacting components on the same organizational level (Andersson and Törnberg, 2018). In complex systems, the components 'are to some degree independent, and thus autonomous in their behaviour, while undergoing various direct and indirect interactions' (Heylighen et al, 2006, 125). The macrodynamics of complex systems emerges through 'self-organization', and by controlling the infrastructure, the outcome of self-organization can effectively be designed. Complexity is the epistemic structure of the digital, and as capitalism is becoming digital, complexity is coming to lie at the epistemic foundation of contemporary modernity.

Conclusion

The powers and capacities of digital platforms first emerged with great expectations – narratives of a 'sharing economy' beyond both market and state, embodied in platforms like Wikipedia and CouchSurfing. Digital technology seemed to bring promises of new social infrastructures that could enable scalable forms of 'commons', and more egalitarian forms of exchange.

Instead of fulfilling such promises, however, we have in recent decades seen digital technology enabling capital to not only further its conquest of commons, but undermining democratic power, weakening public services, promoting labour precarity, violating privacy, and destabilizing the world's democracies. Through code and data, governance is depoliticized and privatized, organized as proprietary algorithms that employ massive behaviour data to engineer social systems through infrastructural design – turning social issues into technical problems to be solved through private means.

The shifts accelerated and made visible by the COVID crisis were already years in the making. Platform corporations like Google and Apple agreed to share with cities and governments a small whiff of their immense trace data – such as the exact location and movements of individuals around the world. News media published agent-based contagion models showing the virus spread in social networks, explaining complexity theory terms like 'feedback loop' or 'non-linearity.' Companies, NGOs, and governments created online dashboards to visualize the data traces of the spread of the virus and the efficacy of government action. Government engagement in such data power was perhaps primarily performative – a Silicon Valley cargo cult built on the aesthetics of Big Tech – as they sought to conceal the effects of decades of neoliberal cutbacks on the emperor's clothing that is public governing capacity.

The word *crisis* comes from the Greek *krisis*, meaning 'decision.' Hippocrates used the word to describe a moment of uncertainty in the progress of a disease, at which the patient arrives at the fork in the road: one path leading to recovery, and the other to death. As the accession of data power has been made visible by the pandemic, we are perhaps finding ourselves at such a choice. We can yet envision digital technology that enables new forms of democratic governance, supporting transnational regulatory governance institutions to face the globalization of capital (Scholz, 2016; Schneider, 2018). Platform infrastructures could be strictly regulated in service of a democratically defined public good, with platform design made subject to political and democratic decision-making. Such models could include what has been referred to as 'platform cooperativism' (Scholz, 2016; Schneider, 2018) or 'platform socialism' (Muldoon, 2022), in which non-market actors are charged to use the capacities of digital technology for the expansion of non-market values such as solidarity, democratic ownership, or seeking to achieve fair labour conditions by, for instance, implementing digital forms of collective bargaining processes.

As Marx and Engels noted, the *decision* inherent in any crisis is by nature a political one. The path that is chosen is determined not by technology, but by political organizing and collective action. To take the road which leads to more democratic and egalitarian ends depends on us exercising our political imagination, and taking charge of events – requiring flexing political muscles that have atrophied during decades of neoliberal hegemony. As data has transformed governance and politics, shaping a future calls on the realization that data is regulation, data is institutions, data is power – and so it must be made subject to the political life of the community.

References

Andersson, C. and Törnberg, P. (2018) 'Wickedness and the anatomy of complexity', *Futures*, 95: 118–38.

Barns, S. (2020) *Platform Urbanism: Negotiating Platform Ecosystems in Connected Cities*. London: Palgrave Macmillan.

Bassens, D. and Hendrikse, R. (2022) 'Asserting Europe's technological sovereignty amid American platform finance: Countering financial sector dependence on Big Tech?', *Political Geography*, 97. https://doi.org/10.1016/j.polgeo.2022.102648

Bauman, Z. (2000) *Liquid Modernity*, Cambridge: Polity Books.

Baumol, W.J. (1986) 'Contestable markets: An uprising in the theory of industry structure', in *Microtheory: Applications and Origins*, Cambridge, MA: MIT Press, pp 40–54.

Bigo, D., Isin, E. and Ruppert, E. (2019). *Data Politics: Worlds, Subjects, Rights*, London: Routledge.

Boyd, D. and Crawford, K. (2012) 'Critical questions for big data: Provocations for a cultural, technological, and scholarly phenomenon', *Information, Communication & Society* 15(5): 662–79.

Castellani, B. (2014) 'Focus: Complexity and the failure of quantitative social science', *Discover Society* 14 [online], available from: https://archive.disc oversociety.org/2014/11/04/focus-complexity-and-the-failure-of-quant itative-social-science/

Cheney-Lippold, J. (2011) 'A new algorithmic identity: Soft biopolitics and the modulation of control', *Theory, Culture & Society*, 28(6): 164–81.

Collier, R.B., Dubal, V.B. and Carter, C.L. (2018) 'Disrupting regulation, regulating disruption: The politics of Uber in the United States', *Perspectives on Politics*, 16(4): 919–37.

Conte, R., Gilbert, N., Bonelli, G., Cioffi-Revilla, C., Deffuant, G., Kertesz, J., Loreto, V., Moat, S., Nadal, J.-P., Sanchez, A., Nowak, A., Flache, A., San Miguel, M. and Helbing, D. (2012) 'Manifesto of computational social science', *European Physical Journal Special Topics*, 214: 325–46. http://dx.doi.org/10.1140/epjst/e2012-01697-8

Couldry, N. and Mejias, U.A. (2019) *The Costs of Connection: How Data Is Colonizing Human Life and Appropriating It for Capitalism*, Stanford, CA: Stanford University Press.

Culpepper, P.D. and Thelen, K. (2020) 'Are we all Amazon Primed? Consumers and the politics of platform power', *Comparative Political Studies*, 53(2): 288–318.

Cusumano, M., Yoffie, D.B. and Gawer, A. (2020) 'The future of platforms', *MIT Sloan Management Review*, [online] 11 February, available from: https://sloanreview.mit.edu/article/the-future-of-platforms/

Davies, H., Goodley, S., Lawrence, F., Lewis, P. and O'Carroll, L. (2022) 'Uber broke laws, duped police and secretly lobbied governments, leak reveals', *The Guardian*, [online] 11 July, available from: https://www.theguardian.com/news/2022/jul/10/uber-files-leak-reveals-global-lobbying-campaign

Érdi, P. (2007) *Complexity Explained*, Cham: Springer.

Foucault, M. (2008) *The Birth of Biopolitics: Lectures at the Collège de France, 1978–1979*, translated by G. Burchell, Cham: Springer.

Harvey, D. (2007) *A Brief History of Neoliberalism*, New York: Oxford University Press.

Hecht, G. (2009) *The Radiance of France: Nuclear Power and National Identity after World War II* (2nd edn), Cambridge, MA: MIT Press.

Hepp, A. (2020) *Deep Mediatization*, London: Routledge.

Heylighen, F., Cilliers, P. and Gershenson, C. (2006) 'Complexity and philosophy', arXiv [online], available from: https://arxiv.org/ftp/cs/papers/0604/0604072.pdf

Isin, E. and Ruppert, E. (2020) 'The birth of sensory power: How a pandemic made it visible?', *Big Data & Society*, 7(2).

Kitchin, R. and Dodge, M. (2011) *Code/Space: Software and Everyday Life*. Cambridge: MIT Press.

Langley, P. and Leyshon, A. (2017) 'Platform capitalism: The intermediation and capitalization of digital economic circulation', *Finance and Society*, 3(1): 11–31.

Larkin, B. (2013) 'The politics and poetics of infrastructure', *Annual Review of Anthropology*, 42: 327–43.

Lazer, D.M.J, Pentland, A., Watts, D.J., Aral, S., Athey, S., Contractor, N., Freelon, D., Gonzalez-Bailon, S., King, G., Margetts, H., Nelson, A., Salganik, M.J., Strohmaier, M., Vespignani, A. and Wagner, C. (2020) 'Computational social science: Obstacles and opportunities', *Science*, 369(6507): 1060–2.

Lohr S (2015) *Data-ism: Inside the Big Data Revolution*, New York: Simon and Schuster.

Mackenzie, A. (2012) 'More parts than elements: How databases multiply', *Environment and Planning D: Society and Space*, 30(2): 335–50.

Malaby, T.M. (2009) *Making Virtual Worlds: Linden Lab and Second Life*, Ithaca, NY: Cornell University Press.

Marres, N. (2017) *Digital Sociology: The Reinvention of Social Research*. Cambridge: Polity Books.

Mason, P. (2016) *Postcapitalism: A Guide to Our Future*, London: Macmillan.

McAfee, A. and Brynjolfsson, E. (2017) *Machine, Platform, Crowd: Harnessing Our Digital Future*, New York: W.W. Norton.

Michod, R.E. and Nedelcu, A.M. (2003) 'On the reorganization of fitness during evolutionary transitions in individuality', *Integrative and Comparative Biology*, 43(1): 64–73.

Morozov, E. (2013) *To Save Everything, Click Here: The Folly of Technological Solutionism*, New York: Public Affairs.

Muldoon, J. (2022) *Platform Socialism: How to Reclaim Our Digital Future from Big Tech*, London: Pluto Press.

Nguyen, C.T. (2020) *Games: Agency as Art*, New York: Oxford University Press.

Nguyen, C.T. (2021) 'How Twitter gamifies communication', in J. Lackey (ed) *Applied Epistemology*, Oxford: University Press Oxford, pp 410–36.

Parker, G., Van Alstyne, M.W. and Jiang, X. (2016) 'Platform ecosystems: How developers invert the firm', Boston University Questrom School of Business Research Paper 2861574. https://dx.doi.org/10.2139/ssrn.2861574

Peck, J. and Phillips, R. (2020) 'The platform conjuncture', *Sociologica*, 14(3): 73–99.

Pfister, D.S. and Yang, M. (2018) 'Five theses on technoliberalism and the networked public sphere', *Communication and the Public*, 3(3): 247–62.

Pollman, E. and Barry, J.M. (2016) 'Regulatory entrepreneurship', *Southern California Law Review*, 90: 383–447.

Rahman, K.S. and Thelen, K. (2019) 'The rise of the platform business model and the transformation of twenty-first-century capitalism', *Politics & Society*, 47(2): 177–204.

Ravenelle, A.J. (2019) *Hustle and Gig: Struggling and Surviving in the Sharing Economy*, Los Angeles: University of California Press.

Rochet, J.-C. and Tirole, J. (2003) 'Platform competition in two-sided markets', *Journal of the European Economic Association*, 1(4): 990–1029.

Rosenblat, A. and Stark, L. (2016) 'Algorithmic labor and information asymmetries: A case study of Uber's drivers', *International Journal of Communication*, 10: 3758–84.

Sadowski, J. (2019) 'When data is capital: Datafication, accumulation, and extraction', *Big Data & Society*, 6(1). https://doi.org/10.1177/2053951718820549

Schneider, N. (2018) 'An internet of ownership: Democratic design for the online economy', *The Sociological Review*, 66(2): 320–40.

Scholz, T. (2016) *Platform Cooperativism: Challenging the Corporate Sharing Economy*. New York: Rosa Luxemburg Stiftung.

Scott, J. (1998) *Seeing Like a State: How Certain Schemes to Improve the Human Condition Have Failed*, New York: Yale University Press.

Srnicek, N. (2016) *Platform Capitalism*, Cambridge: Polity.

Srnicek, N. (2017) 'The challenges of platform capitalism: Understanding the logic of a new business model', *Juncture*, 23(4): 254–57.

Thelen, K. (2018) 'Regulating Uber: The politics of the platform economy in Europe and the United States', *Perspectives on Politics*, 16(4): 938–53.

Törnberg, P. (2021) 'Short-term rental platforms: Home-sharing or sharewashed neoliberalism?' in T. Sigler and J. Corcoran (eds) *A Modern Guide to the Urban Sharing Economy*, New York: Edward Elgar Publishing, pp 72–86.

Törnberg, P. and Törnberg, A. (2018) 'The limits of computation: A philosophical critique of contemporary Big Data research', *Big Data & Society*, 5(2). https://doi.org/10.1177/2053951718811843

Törnberg, P. and Uitermark, J. (2021a) 'Towards a heterodox computational social science', *Big Data & Society*, 8(2). https://doi.org/10.1177/205395 17211047725

Törnberg, P. and Uitermark, J. (2021b) 'Tweeting ourselves to death: The cultural logic of digital capitalism', *Media, Culture & Society*, 44(3): 574–90.

Uprichard, E. (2013) 'Describing description (and keeping causality): The case of academic articles on food and eating', *Sociology*, 47(2): 368–82.

van Dijck, J. (2014) 'Datafication, dataism and dataveillance: Big Data between scientific paradigm and ideology', *Surveillance & Society*, 12(2): 197–208.

van Dijck, J. and Poell, T. (2013) 'Understanding social media logic', *Media and Communication*, 1(1): 2–14.

van Dijck, J., Poell, T. and de Waal, M. (2018) *The Platform Society: Public Values in a Connective World*, Oxford: Oxford University Press.

van Doorn, N. (2019) 'A new institution on the block: On platform urbanism and Airbnb citizenship', *New Media & Society*, 22(10): 1808–26.

Yeung, K. (2017) 'Hypernudge': Big Data as a mode of regulation by design', *Information, Communication & Society*, 20(1): 118–136.

Zuboff, S. (2019) *The Age of Surveillance Capitalism: The Fight for a Human Future at the New Frontier of Power*, London: Profile.

4

Data Ethics in Practice: Rethinking Scale, Trust, and Autonomy

Alison Powell

Introduction

Relentless unfolding of surveillance architectures and an embedding of data exploitation into the foundations of capitalism suggest that data ethics are urgent and necessary. Frequently, however, data ethics refers to or resolves into vague statements of principles by powerful entities: Google's publication of an ethics charter in 2018 and its launch of an Advanced Technology External Advisory Council (ATEAC) – popularly known as the 'ethics board' – in 2019 which lasted only a week before being abandoned (Walker, 2019) are just two examples. Equally, data ethics can refer to design requirements that are presented as an idealized aim for designers of data-based technologies. These include requirements proceeding from regulatory frameworks such as the European Union's General Data Protection Regulation (GDPR) or similar legislation in place in other countries. In this mode, data ethics are often associated with rules that must be followed or consequences that must be managed (Powell et al, 2022). Within this framing, data ethics is often displaced. This displacement can occur temporally, as when adherence to ethical principles is pushed later in time until after a data collection process is completed, or functionally, as when data ethics are perceived as issues of compliance (Powell et al, 2022). As a counterpoint, some researchers are beginning to consider ethics as part of social practice, shifting from discussions of ideal ethical principles that should be addressed in technology design towards discussions of specific contexts and practices (Møller et al, 2020).

Ethics as practice in data-driven contexts refers to ways of organizing, acting with, relating to, or contesting data. The use of data within urban settings provides a number of specific contexts and practices, intersecting and transcending what might be considered 'top-down' or 'bottom-up' dynamics. Data-based governance, management, and civic engagement are deeply embedded into the function and experience of cities, as the other chapters in this book illustrate. This embedding raises important questions of ethics, justice, and power. Regulatory responses, such as data protection legislation and limits on data collection, address some of the most obvious power differentials but cannot necessarily address issues of systemic injustice. This is often because, in contrast to well-specified issues of regulatory compliance, issues of justice are temporally dispersed and contextually specific. This means that 'bottom-up' data ethics practices are embedded within techno-systemic frames maintained through state and corporate narratives, investment, and policy support (Powell, 2021). This does not mean that bottom-up ethical practice is impossible, or only legitimate if it offers straightforward, effective resistance to 'top-down' dynamics. Rather, it suggests that such practice might be most effective when it acknowledges and operates in relation to techno-systemic frames. This means looking at how data-based structures create and maintain unequal power relations, as well as how attempts to intervene in these relations, generate new potential for change as well as new complexities. This approach resists the urge to frame attempts to escape a data-driven universe as ideal ethical positions, and instead attends to the tensions that inevitably emerge in the ways that alternatives to surveillance, extractivism, exploitation, and data profiteering are designed. Through attention to these tensions, new possibilities can sometimes become apparent. Therefore, in this chapter, data ethics is understood as a range of practices that attempt to address issues of justice and consequence related to the design and operation of data-based systems.

This chapter outlines a range of possibilities for understanding issues of data justice from the perspective of ethics as a practice. The practices include commercial practices, which sometimes show the limits of existing regulatory frames, as well as participatory processes like data walking, which can be used as an alternative to standard processes of consultation in urban planning, and the creation of collective models of reflection on the use of specific data – sometimes called neighbourhood ethics committees. These practices model different kinds of engagements with knowledge, data, and with different dynamics of resistance, resilience, and community strength. This makes these practices useful and important ways of understanding the complex dynamics that make up the ethical terrain of smart cities, which I define as urban realms managed at scale with conflicting strands of data and negotiated through a range of knowledge. The chapter therefore reflects on how the processes of trust and autonomy modelled through such practices of ethics might connect with other considerations that apply at different scales.

A reflection on practice and scale is especially important in a context of uncertainty or 'perpetual crisis'. Many aspects of the current ongoing crisis are experienced at small or lived experiential scales through bodily perception, while only being able to be experienced at a global or distributed scale through data and narratives based on data. Thus the climate crisis, which differentially impacts individual bodies located differentially in space, as well as crises of health, institutional resilience, and inequality. This illustrates the paradoxical connections between data and ethical practice: while some aspects of lived experience escape complete datafication, other features of lived experience, including important shared experiences such as the planetary experience of climate change, are intensely datafied. Therefore, ethics as practice, especially in smart city contexts, needs to cut across and connect different scales and aspects of datafication. There is no singular 'data ethics' – nor can data ethics merely reject datafication.

Conflicts of scale and urban smartness

As Cinnamon points out in this volume (Chapter 10), cities become datafied in part through appeals to scale. A combination of the capacity for quantification and the desire to manage large-scale, often complex systems has meant that smart urbanism operates through a scalar politics. This is similar to the way that other technological politics were positioned in the past, including claims that the expansion of the internet separated a purported 'global flow' away from an experience of 'local place' (Castells, 2020). This scalar distinction, where the small-scale and the large-scale are both separated and differently positioned in relation to technological capacity, continues. In relation to datafication and smart urbanism, 'small-scale' projects often assume legitimacy based on an assumption that context is easier to understand at a small scale, by drawing on qualitative rather than quantitative data, or identifying how expanding scales can cause harm by removing context, flattening difference, or intensifying inequalities by embedding biases in large-scale data systems. Creating oppositions between scales and linking these oppositions to competing interpretations of data allows these tensions to become the motor of contentious urban politics, as Cinnamon discusses. At a citywide scale, the assumption that broad-scale data contributes straightforwardly to optimization of urban service delivery (such as traffic management or allocation of assumedly scarce resources) reiterates a 'techno-systemic frame' (Powell, 2021) that foregrounds quantitative data production and analysis as the best way to understand urban life. Such a frame is not inevitably connected with large-scale data politics: grassroots actors also operate within these frames, shifting their civic actions towards engagement with data in order to bolster their legitimacy. What results is often friction and tension: conflict regarding both the meaning of data and also its reliability, validity, and appropriateness as a technique for gaining civic voice.

Therefore, issues of scale intersect with broader dynamics of datafication within smart cities, resulting in frictions at, within, and between scales. This dynamic complicates efforts to oppose, transcend, or transform urban spatiality through recourse to data either as material or as a discourse embedded within techno-systemic frames. Redress of these frictions and the broader injustices or inequalities they reflect becomes an ethical imperative. This is the space occupied by forms of data activism, which can both attend to and leverage fictions in either reactive or proactive modes (Milan and van der Velden, 2016). Data activism can be one form of ethical practice in relation to data, and like other ethical practices, can unfold at and across different scales. Data activism is also constrained by techno-systemic frames, encouraging citizens to present their concerns using data or data-based arguments. This can contribute to the failure of purportedly 'ethical' projects to address systemic issues.

Data ethics, regulation, and practice

Within policy and regulatory spaces, the phrase 'data ethics' has already been captured by powerful actors who use it to suggest the legitimacy of their existing business models and to disavow the necessity to develop or abide by laws or regulations. The establishment of ethics committees within monopoly capitalist data firms and the production of ethical codes of conduct are examples of this performative 'data ethics'. In response, scholars and practitioners are increasingly turning towards an examination of ethics as practice, that encompasses actions such as organizing using data or contesting data meaning and power. Ethics as practice also focuses on what is done, rather than what is said. In an age of perpetual crisis, it is actions that shape the space of engagement, and both powerful and less powerful actors do things that run counter to, or open different space from, what is declared. In other words, in crisis, plans give way to situated actions (see Suchman, 2007). Responsiveness, in a crisis state, may mean ignoring or bypassing the regulatory frameworks that are often indicated as the foundations on which ethical action might be taken. Equally, it may also involve actions that create new or unexpected ways of understanding or acting with data.

For example, regulations already govern many aspects of data collection, use, and processing at large scale. These include wide-scale regulations like the GDPR which applies across the European Union and which also influences policy in the UK, as well as data processing and procurement rules, which apply in specific sectors. Regulatory frameworks, ideally, form the foundation for actions or practices that address substantive issues of justice and equality. They also, however, act as terrains of struggle where issues of trust, autonomy, and context are brought into focus.

In this chapter, examples of data ethics in practice that operate at different scales provide an indication of how practices of data ethics unfold within the UK's peculiar form of perpetual crisis (featuring a lax and inequitable response to COVID-19, Brexit, a weak regulatory environment, intense and racialized inequality, and a political incapacity to address climate change). At the national scale, regulatory gaps created by COVID-19 emergency legislation have reopened questions about the collective value of data, the potential or limits of trust in different kinds of institutions, the role of civil society organizations in performing data activism, and the limited potential to shift data governance frameworks. At a hyper-local scale, similar issues of trust, autonomy, and context emerge around the potential to create local groups of citizens tasked with creating data management strategies that align with local conceptions of value, fairness, and justice. Between these two scales lies the potential to investigate ethics as practice as a means to surface other forms of knowledge and care that might be necessary for a flourishing existence in a state of perpetual crisis. These practices might include research practices like data walking, or experiments in creating multiscalar relational structures that allow for dynamics of mutual aid and support to proliferate.

Large-scale (un)ethical practices

Through the early stages of the COVID-19 pandemic, the back-end delivery of health care in the UK became quietly intermediated by companies invested in using AI technologies to dynamically manage health care resources. In April 2020, regular procurement rules that would usually have been in force to regulate the process for awarding tender contracts for government services were suspended under emergency legislation passed in order to deliver personal protective equipment (PPE) that was in short supply in the UK. Palantir, a US-based data management company whose previous core business included managing data for the United States intelligence services and for its Immigration and Customs Enforcement system, won a contract to manage data during the pandemic. This bid was awarded at an artificially low cost, suggesting that Palantir was seeking to make its systems part of the UK health infrastructure. In 2021 Palantir was removed from a UK government health and social care contract after public outcry, facilitated in part by civil society organization Foxglove, which pointed out that Palantir's move from US-based intelligence technologies to UK health systems was facilitated at least in part by the suspension of the UK government's public procurement rules during the COVID-19 pandemic and the sense of a latent and unrealized benefit of health data. In 2022, Palantir submitted a bid to take responsibility for end-to-end management of the UK's health infrastructure, which Foxglove continues to oppose.

Foxglove's actions identify ethical conflicts unfolding at very broad scale, drawing attention to regulatory gaps and also to the fact that the creation of a national-level health service and associated data infrastructure have generated a collective store of beneficial data which can be managed in a number of different ways. The possibility for health data to become a store of shared value and benefit contrasts with the techno-systemic frame that positions it as input data for resourcing and policing algorithms that might be sold on in other contexts. Cori Crider of Foxglove writes, 'If the future of UK health and social care depends on better data, a sustainable system needs to build up our own data science expertise, and not put us in hock to expensive consultants and tech firms' (Crider, 2021). Foxglove's opposition to Palantir's role in the NHS identifies how processing and managing this extremely broad-scale data has long-term benefit for the processor, and also the fact that existing regulatory frameworks do not make adequate provision for the loss of collective value to the UK as a whole when this data is moved away. Foxglove's assessments of the risks to the UK mirror some earlier data activism regarding health data, which appeared as opposition, in 2011, to a data-sharing agreement called Care.data, which made it much easier to share patient data between hospitals and family doctors, but which also would have opened this data to reuse by private insurance companies (Carter et al, 2015).

The quiet and systemic intervention by Palantir raises some questions about the ethics of AI in health delivery and health resource management, as well as broader questions regarding the ethical practices at work in the programming, marketing, and global reach of data-based systems. Palantir's resourcing systems, for example, will have been trained on healthcare data from the United States, set in a context of extreme health inequality, where poor and minoritized communities lack effective access to healthcare, meaning that systems (trained with the data that is available) effectively fail to include these communities. This is one of two types of bias associated with these kinds of data-driven systems: the bias of exclusion. The other type of bias, the bias of inclusion, results from the ways that assumptions about minoritized communities are 'programmed in' to such systems, leading to the continuation of racist or discriminatory practices; for example, in relation to chronic conditions such as diabetes that can be exacerbated by poverty or inequality. In addition, in a public system like the UK's health system, is it right for the insights from these data to accrue to a private entity like Palantir? Here we encounter a problem of data governance and collective management of data. These data are valuable – UK health data is particularly valuable – and they are public assets through a certain definition. There is strong public support for retaining the value of this data, which could, with the right kind of political or economic argument, be leveraged as a form of collective benefit.

There is also another aspect of scale and autonomy that emerges in the case of Palantir and the NHS data, which connects more strongly with the ways that scale, data, and autonomy are positioned in smart cities. This has to do with the way that the management of data, including the way that it is processed, managed, and placed into dashboards, establishes the power of the intermediary. In UK smart cities, data processing contracts and dashboarding services are often also awarded to multinational corporations such as Siemens, which facilitates many location-based sensing systems (Siemens, 2020). With access to large-scale data like NHS medical data or smart-city mobility data, data processing intermediaries like Palantir or Siemens construct dashboards, which can then be marketed back to clients like cities as the main means to understand broad-scale phenomena. What's at stake in a dashboard-driven world are the specific ways that dashboards create and represent the truth. Scholar of dashboards and their histories Nate Tkacz writes, 'Dashboards condense data for easy digestion, which can obscure a user's knowledge of how trustworthy or accurate that data is. By presenting often very complex, messy and varied data in simplified forms for consumption via a dashboard, sometimes subtle changes take place in how that data is understood' (Bartlett and Tkacz, 2017). The power to shape that data lies with the creator of the dashboard, who sets the terms through which certain things are defined or processed as data, as well as the way that data are standardized, codified, and managed over the long term.

Trust and autonomy at and across scale

These questions of scale and power reinforce the idea that trust is undermined and autonomy is eroded when data-based decisions are taken at a broad scale, meaning that protection of individual rights needs to be delegated to regulatory frameworks. However, as illustrated by the Palantir case, even existing regulatory frameworks may be suspended or not correctly applied, which may be one reason why trust in data-based systems has become more fragile. However, large-scale data processing, especially of complex urban data like mobility data, can also reveal complex, localized patterns, incongruities, and complexities. Batty (2022) identifies the need to interpret and understand this data using principles of relationality, marking a shift away from assuming that large-scale data generate highly generalized insights. Batty's team created speculative models of different kinds of mobility patterns that might be predicted for a post-pandemic lockdown London based on different combinations of choices, such as working at home, working in an office, or prioritizing different forms of transit. The simulations of different scenarios for 'post-lockdown' urbanism were more specifically various than the authors had anticipated. This suggests that the techno-systemic frame foregrounding computable and dashboardable data's seamlessness and consistency may embed more frictions than expected. Batty's revision of the kinds of insights generated from large-scale data suggests that autonomy and relationship complicate predictions made using large-scale data sets. Therefore, the practices of dashboarding and predictive data analysis create and maintain dynamics where individual behaviour is observable and interpretable by corporations whose responsibilities to individuals and citizens may be limited, ignored, or unable to be fully considered. This creates tensions that are difficult to investigate, in part because they cut across and between scales.

Taking forward this realization of complexity and friction even within 'broad-scale' data collection reiterates the importance of regulatory protections. It also indicates that deeper understandings are needed of how people understand what aspects of their lives are rendered into 'large-scale' data and what significance this might have. These ethical questions are, once again, framed or described in relation to scale, where 'smaller-scale' frames of assessment suggest the capacity for attention to qualitative, rather than quantitative or AI-processed aspects of data. However, scalar oppositions don't always map straightforwardly to questions of interpretation. What might be more valuable than celebrating the 'small scale' for its own sake could be the foregrounding of experience and complexity, which is also in evidence in Batty's re-examination of large-scale data.

Small-scale interventions and shifts in practice create opportunities to reveal, unfold, and contest the dynamic of data-based smart urbanism that Kitchin (Chapter 2, this volume) describes, whereby dynamics of capitalist extraction intensify the power of commercial companies within urban governance processes. As Cohen (2019) points out and Kitchin (Chapter 2) develops, the dynamics that result from this consolidation of economic power, data extraction capacities, and control of governance processes by aligning them with data-driven decision making, changes the expectations and performances of citizenship (Powell, 2021).

Many civic actions are now undertaken by, through, and in relation to the data-driven dynamics that characterize the smart city. These can include active modes of data citizenship such as data audits of open government data, and civic projects such as environmental sensing, map-making, or 'bottom-up' data advocacy (Couldry and Powell, 2014; Gabrys, 2016). They can also include socio-technical efforts that are less explicitly technology-driven, including efforts to shape and reframe how different kinds of knowledge might connect, or contest, digital data. This socio-technical version of 'data friction' highlights how urban data power is neither a matter of total domination through commodification and surveillance dynamics, nor is it a matter only of opposition through resistant data power. Instead, the qualities of social friction and tension that emerge around the practices of data collection, the definitions of which knowledge is valuable and important in relation to this data, and the storage, sharing, management, and brokerage of this data create the conditions for emerging forms of solidarity.

Reconfiguring the value of data and creating new solidarities

For example, frictions concerning the quality and use of sensor data intended to facilitate community-based reflection, decision-making and communication between a neighbourhood where poor quality accommodation suffered from damp emerged when the community of Knowle West in Bristol undertook a pilot study using community-collected sensor data (Balestrini et al, 2017). The repository of this data was intended to become a 'commons' for use among a community comprised of social housing residents, tenants in privately-owned buildings, small business owners interested in potentially creating businesses around the collection of data about building conditions, and the local government, whose austerity-driven cost-cutting resulted in the firing of the inspector who previously judged the difference between damp and humidity. These multiple groups each had different ideas of how and in which way to interpret and use sensor data, especially in a context where there were asymmetries in terms of the ability to make data make money, secure power, or change institutional knowledge. The frictions and tensions that emerged around the data commons produced a form of solidarity between technical experts, community leaders, and residents in poor quality housing that allowed them to identify the gaps in knowledge, expertise, and action. The frictions necessitated a practice of solidarity, grounded in a shared recognition of the gaps in capacity at the government level.

These gaps in capacity have led to considerations of how to embed data into work that strengthens the capacity of different kinds of organizations, including informal organizations, to support public services and, more importantly, the public. The UK's Ratio Research conducts research in partnership with public services and community support organizations investigating connection, trust, and belonging as a foundation for public services and civil society capacity. Michael Little describes this as 'relational social policy' (Little, 2021). One of Ratio's long-term research endeavours has been to create opportunities for connection, trust, and belonging to develop by setting up collective contexts where resources, including data, are shared and employed for mutual benefit. One of these contexts is the 'data club', a version of a mutual aid group that reflects on the potential to use data to enhance trust and connection.

Data clubs and local ethics boards

In Glasgow, Rotterdam, Birmingham, and London, small groups have formed based on principles of mutual aid. In Glasgow, groups of women are supported to begin saving money together. Through the process of making small loans and saving together, the women create relationships of trust and reciprocity, able to talk about their struggles and share strategies in response. The groups have also begun to experiment with sharing data collected from wearables and apps, using these data to create conversations and provide structures that allow the women to feel safe with one another. One of the features of sharing data in these contexts is to unseat feelings of shame that result from individuals internalizing high levels of stress and operating without strong relationships of mutual support. Creating these 'data clubs' lets women read their data in relation to others', setting up emotional norms within the group that support the capacity to discuss difficulties – which could even appear 'within the data' as divergences or outliers. Data clubs allow their participants to define or redefine the shared values they might hold and what their data might mean for them. The practices within data clubs currently include sharing, discussing, and comparing, although it is possible that defining shared value might also include defining the conditions under which data might be shared or sold, including where any resulting benefit might accrue. The data clubs show the potential for community power to enfold data practices into other efforts to establish and maintain relationships of trust and reciprocity. Yet community institutions can also provide capacity for extending the mutual aid principles of data clubs beyond a small group, at the same time modelling different interscalar dynamics for supporting dynamics of mutual aid.

One of the communities is in Walworth, south London, a neighbourhood of 40,000 people where 83% of the population is in the most deprived quintile nationally and where 52% of the population is of Black or minority ethnic background. I have joined Ratio Research as a community-based researcher exploring what data East Walworth residents think is important, what they would like more data about, and what questions their community could answer with data, in preparation for creating a set of data clubs in the local area. In Walworth there could be up to 15 data clubs of 20 people each. Data club members would receive £70 per year into the club bank account, be expected to keep in touch by WhatsApp, save between £1 and £5 per week in the club bank account, and collect data on community and health. A club member would also be expected to meet every two weeks with the members to reflect on what their data mean for the group and the broader community, and to make loans to other club members. The structure and practices of the clubs draw direct inspiration from other mutual aid structures, particularly the trust and safety net created through the provision of small loans.

To link together the Walworth data clubs and to help to consider how data might create new relationships of trust locally, beyond the small scale of the individual groups, the clubs will be supported by a neighbourhood ethics committee. The committee should provide a point of connection between

Figure 4.1: The 'data stall' at a Walworth community event

the individual clubs as well as helping to reflect on and design mechanisms for the collection, governance, and perhaps sharing of this data.

To support the creation of a neighbourhood ethics committee, we spoke to some East Walworth residents, primarily from a few blocks around a single street, about what they thought data might mean or do for them (Figures 4.1 and 4.2). We discovered that residents are interested in data, which for them extends to inclusion of people who don't use the internet into what is perceived as a data-driven culture or economy. We also discovered that when residents talk about data they talk about connection. They want to know what is going on in the area in which they live, including who lives in the area and whether they have things in common (for example, mums connecting to mums), what is going on their area, and information that might be specifically local, such as the location of accessible green spaces. Many residents linked data to storytelling, wanting both to share and to know more about the history of their area. Residents talked about wanting to use data to influence local 'issues' that they found important or politically significant, including loss of green spaces, waste, climate, and the quality of local services. We discovered that, similar to the way that data clubs allow for practical as well as emotional reflections and forms of support, Walworth residents were interested in using data that developed hyper-local services, or that could support their everyday lives by making scarce local resources more accessible or more fairly distributed. However, they also wanted to

Figure 4.2: Soliciting ideas and conversations about what data is needed, important, or at stake within the Walworth community

share and receive information that could strengthen identity or belonging. They were honest about the fact that existing data-based services such as the Nextdoor app didn't quite meet these needs, and nor did information from the local government.

What this ongoing research suggests is that within the frameworks provided by mutual aid, different kinds of interrogations and explorations of data might be possible. Data might simultaneously be a resource to draw on in much the same way as shared savings in a mutual aid club, or it might facilitate political action by highlighting tensions between residents' experiences and data or information collected by the local government. It is possible to imagine different kinds of data practices being supported, interrogated, or planned through the contributions of a neighbourhood ethics committee to the practices of data clubs. For example, the committee might support the community to develop different kinds of data-based products and services, or to collect data about local experiences that could open a difficult conversation with the local government, in much the same way as occurred in Knowle West through the data commons that defined damp as a political issue. The ethics committee might also think carefully about how to facilitate relationships *between* data clubs and therefore to knit together different configurations of connection, trust, and autonomy. A neighbourhood ethics committee process thus creates an opportunity for a critical praxis of data ethics: including the capacity to define how to produce, share, manage, limit, mitigate, or otherwise reimagine data use about, or in service of community members. The significance of these efforts is not in their potential to scale up structures of mutual aid, nor to scale the value of data by collecting more of it. Instead, the significance comes from the capacity for data to be a carrier of relationships, of value, and of collective autonomy. This aligns efforts to sustain forms of mutual aid and relational care across and between scales, which also creates some of the conditions required to sustain social relations in crisis (Harrington and Cole, 2022).

Data-driven experience transcends scale

The examples in this chapter have shown how ethics in practice in relation to data challenge and transcend separations of 'large' and 'small scales'. They also illustrate that what's most often at stake in data-mediated relationships are issues of trust, connection, and autonomy. These issues are represented in varying ways in and through data and data-based systems. Balancing and transcending separations of scale requires other means of investigating how data come to mean things, and how different or more complex notions of trust, autonomy, or relationality might be developed. One mechanism for investigating how these principles might be experienced in practice could come through structured, collective experience. The 'data walking' methodology I have developed and iterated creates opportunities to practise and explore different forms of urban knowledge creation (Powell, 2018a; 2018b). In particular, data walking using role-playing and a processual experience of moving through and observing urban spaces while playing specific and well-defined roles of 'navigator', 'observer', 'interviewer', 'map-maker', and 'photographer'. Part of the delight of a data walk is the experience of needing, as a group, to define and operationalize what 'data' means, and to do this while playing the assigned roles. Data walking, as a research and public engagement process, seeks to reveal and reconfigure hierarchies of urban knowledge production. By beginning a collective, observational process of movement through an urban setting by *defining data*, data walking invites participants to acknowledge that digital data or data-processing infrastructures are only one type of urban data. Furthermore, the fact that data walking requires and embeds performative methods and role-playing as either a note-taker, map-maker, observer, interviewer, or photographer reiterates how different kinds of knowledge combine to define and shape 'what's counted' and 'what counts.' Urban knowledge, as Shannon Mattern (2021) points out, can't be reduced to a computational figuring of data needing to be processed. Data walking, in the way I have practised it, attempts to resist this figuring. What data walking means to do is not to excavate the city straightforwardly, but rather to provide experiences of how data are brought into being. Etymologically, data are 'what is given' (see Rosenberg, 2013), but not all data are given to all. Rarely are urban residents provided the opportunity in everyday life to perform the expertise that surfaces, defines, and renders aspects of urban life that are 'in the background' into data that are 'in the foreground.' This creative research practice creates an opportunity for people to experience the making of data, including the way that this depends on the interactions between different performances of expertise and different professional identities.

Between scales: friction, tension, and turbulence

An expansive data ethics that is positioned within the frame of data-driven smart cities would need to acknowledge the paradoxes of intensive datafication and the ways that these cut across scale: the fact that on one hand, data extraction acts as a means to abstract and control individual experience; and that on the other, data-driven systems provide the ability to understand and position urban experience as something more than individual. These paradoxes are more significant in light of experiences of perpetual crisis. An example here is of climate change. As Paul Edwards (2010) masterfully illustrates, the very notion of climate as something that can be described or experienced on a global scale depends on collection, maintenance, and interpretation of data. Therefore, individual and situated experiences of extreme weather can be positioned as shared experiences of a changing climate. In their work on data ethics using the language of 'data feminism', Catherine D'Ignazio and Lauren Klein (2020) connect this situated, yet global experience to situated objectivity as sketched out by Donna Haraway in her discussions of the benefits of partial perspectives and the contestation of a singular point of objectivity.

In relation to data ethics in practice in urban settings, these insights inspire attention to the ways that situated knowledge or situated objectivity connect with or engender forms of solidarity forged in difference. Once again, this notion of situatedness and solidarity complicates the idea that data power is built through the conflict between extraction and modelling at the large scale, and specific and contexualized experience at small scale. Recent work on friction and tension in relation to data, knowledge, and practice has described the ways that conflicts regarding the use of data reveal gaps and contingencies of local knowledge and the potential – and limits – of data in addressing these (Powell, 2021), These ideas have also been used to illustrate the asymmetries and contingencies within global processes of data circulation (Lehuedé, 2022). Lehuedé argues that 'data turbulence' begins to apply at a global scale as a result of the increase in friction, leading to phenomena such as competitive cybersecurity efforts. Both of these views of tension draw from Anna Tsing's identification that friction is 'a central feature of all social mobilizing [...] based on negotiating more or less recognized differences in the goals, objectives, and strategies of the cause' (Tsing, 2005, xii). She argues that misunderstandings within long-term social movements, far from producing conflict, actually permit people to work together. Tsing uses friction to understand how heterogeneous, unequal encounters produce energy, questioning the inevitability of seamless global flows.

Building from this, data turbulence highlights that the dynamics of tension and relationship between entities are not easy to reduce to opposing issues of 'globality' or 'locality' or even to 'materiality' and 'sociality.' Instead, the turbulence of data-based relationships playing out across the globe involves both material infrastructure as well as political philosophies. Lehuedé argues that attention to this discursive aspect highlights, in particular, the antagonism that characterizes different interpretations of digital sovereignty – including the question of whether such sovereignty is indeed possible. In the context of data-based systems, turbulence impacts questions of data localization, data storage, and the disparate material and economic impact of data processing on different geographies. This is especially pertinent when considering the climate crisis. Here, data processing displaces the long-term risks and harms of data processing into fragile geographies sometimes distant from where data are collected.

These complexities of turbulence illuminate how trust and autonomy may be decoupled from scale. While Lehuedé's work identifies how autonomy allows for a connection with territory and the claims to knowledge, experience, and legitimacy that stem from a strong connection to territory, this does not mean that these claims apply only at a small scale. Rather, data turbulence characterizes how materiality and meaning both cut across large and small scales, unsettling governance arrangements as well as claims on meaning.

This notion of a zone of turbulence encompassing both local tensions and global material infrastructures helps to break down rigidity in interpretations of scale. While it is true that 'all data are local' (Loukissas, 2019) in the sense of being most meaningful when they are closest to the situations and knowledge through which they have been defined, it is not true that data are only meaningful or interpretable at small scale. As I described above, the practices of data intermediaries working at broad scales create different kinds of conflicts that are related to phenomena beyond the decontextualization, aggregation, or 'delocalization' of data.

Once we begin to see scale as turbulent within, across, and between the 'global' and the 'local', it becomes possible to examine the way that trust and autonomy are already challenged by extensions in space and time by different models of ownership and control of data. Following this, we can begin to see how these models might be attenuated or reconfigured across different scales and temporalities.

Here, it is useful to draw on design theorist Arturo Escobar's notion of a 'pluriverse' – a 'world where many worlds fit' (2018). A pluriversal orientation to practices of data ethics would recognize that the world of the Knowle West residents and their spreadsheets of damp data, or the residents of council blocks in East Walworth and their local histories, can exist within and alongside a concern over the kinds of information that is, or is not, shared by the local government or the national state. Equally, the pluriverse can contain a world in which East Walworth residents convene a neighbourhood ethics committee that shifts the kind of products and services designed for and used by its residents. Escobar's concept resists the idea that all things need to be related to each other along a single logical axis, reminding designers and practitioners that different feelings, conceptions, beliefs, and actions can exist simultaneously, shaping how different experiences and outcomes can result from encounters with the same things. If this world seems constrained by the current political-economic realities, that does not make it impossible.

Indeed, a focus on friction reinforces the way that different possibilities emerge as a result of tensions or difficulties. Anticipating difficulty rather than seamlessness provides a way for widely differing practices to become apparent, challenging the notion of a smooth shift in scale from small to large or a seamless set of data predictions. Friction may be inherent in commons. Resource-based commons, like pasture or fishing grounds, are governed by agreements between all of the beneficiaries and participants in the commons. Often, researchers assume that the management of these commons depends on applications of consistent rules within relatively small, homogeneous communities (Hess and Ostrom, 2005). This isn't always the case. In investigations of data commons, for example, researchers contrast the way that expansions of intellectual property create 'anticommons' that produce conflictual relationships around the use of data, with the production and management of data commons providing an alternative (Fisher and Fortmann, 2010). Writing about the ethics of contributions to data resources like biobanks, Prainsack and Buyx (2017) argue that solidarity is expressed through action, not thought. In Knowle West, for example, there was little sense of a shared position or problem until sensing data started being collected, at which point 'data defined the community [...] defined the way people thought about community' (Powell, 2021). The data collection thus established action.

Conclusion: Transgressing scales as a matter of structural care

This chapter has explored how frictions evident at, between, and transgressing different scales of 'smart' or 'data-driven' systems highlight the importance of trust and autonomy in undoing dynamics of exploitation, extractivism, or alienation. It has also identified that creating opportunities to decide on what data is and how it should be made, managed, and valued can unseat assumptions about the inevitable, perpetually enduring power of large-scale data processors. This is not to discount the practices that the chapter also outlines, whereby already powerful actors create or employ regulatory gaps in order to create new means of commodifying or profiting from data. All of these practices are dynamic, and all are absorbed in creating and defining meaning and value. In addition, they take place within a context of perpetual crisis, where techno-systemic frames are often leveraged as means of survival, suggesting a lack of alternatives to broad-scale, abstracted, and alienated modes of governance. Yet, as always, other possible worlds can be contained in this one. To conclude, I reflect on how the unruly scales of practice presented in this chapter might demonstrate structural care.

Mattern (2021) describes the importance of embedding understandings of maintenance and repair into considerations of urban data ethics, connecting these to other contemporary struggles, 'Amid the uprisings of spring and summer 2020', she writes, 'Deva Woodly emphasized that #BlackLivesMatter is a movement committed to *structural*, rather than merely individual, care' (p 120). Care is contentious: as Judith Butler (2020) points out, it is not always consensual. It can also be enacted as compulsory within some professions and (often gendered) social roles: care is expected across the service industry and especially by nurses, social workers, and receptionists, as well as by day care and nursery workers – not to mention parents. Care links together experiences of both autonomy and vulnerability, and hinges on trust. We can see care expressed across the frictions and tensions, the turbulence and the incommensurability of data. Structurally, care is enacted in establishing regulatory frames, but perhaps even more significantly, in ensuring that these are enforced. At the level of service delivery, the concept of 'relational public services' calls on the practices of care that are the legacy of charitable and voluntary organizations, while the frictions that emerge around the use of data in these contexts demand new forms of expertise, participation, and civic attentiveness. These are all new forms of care, of course.

The question, as the perpetual crisis continues to shrink states, delegate responsibility for urban systems to powerful corporations, and to reduce the liberal structures for democratic participation, is how to exist within the turbulence, how to enhance practices and infrastructures of care, and how to imagine other possible futures. If this chapter has done some of this, I give thanks.

Acknowledgements

Some material in this chapter has been developed in collaboration with Michael Little of Ratio Research. Portions of the chapter are based on work published in *Undoing Optimization: Civic Action and Smart Cities* and on work undertaken with the Ada Lovelace Institute. Thank you to Adam Greenfield for useful discussions on mutual aid.

References

Balestrini, M., Rogers, Y., Hassan, C., Creus J., King, M. and Marshall, P. (2017) 'A city in common: A framework to orchestrate large-scale citizen engagement around urban issues', in *Proceedings of the 2017 CHI Conference on Human Factors in Computing Systems*, pp 2282–94.

Bartlett, J. and Tkacz, N. (2017) 'Governance by dashboard: A policy paper', London: Demos.

Batty, M. (2022) 'The post-pandemic city: Speculation through simulation', *Cities*, 124. https://doi.org/10.1016/j.cities.2022.103594

Butler, J. (2020) *The Force of Non-Violence: An Ethico-Political Bind*, London: Verso.

Castells, M. (2020) 'Space of flows, space of places: Materials for a theory of urbanism in the information age', in R.T. LeGates and F. Stout (eds) *The City Reader* (7th edn), London: Routledge, pp 240–51.

Carter, P., Laurie, G.T. and Dixon-Woods, M. (2015) 'The social licence for research: Why *care.data* ran into trouble', *Journal of Medical Ethics*, 41(5): 404–9.

Cohen, J.E. (2019) *Between Truth and Power: The Legal Constructions of Informational Capitalism*, Oxford: Oxford University Press.

Couldry, N. and Powell, A. (2014) 'Big data from the bottom up', *Big Data & Society*, 1(2). https://doi.org/10.1177/2053951714539277

Crider, C. (2021) 'The UK government has ended Palantir's NHS data deal. But the fight isn't over', OpenDemocracy, [online] 15 September, available from: https://www.opendemocracy.net/en/ournhs/the-uk-government-has-ended-palantirs-nhs-data-deal-but-the-fight-isnt-over/

D'Ignazio, C. and Klein, L.F. (2020). *Data Feminism*, Cambridge, MA: MIT Press.

Gabrys, J. (2016) *Program Earth: Environmental Sensing Technology and the Making of a Computational Planet*, Minneapolis: University of Minnesota Press.

Harrington, E. and Cole, A. (2022) 'Typologies of mutual aid in climate resilience: Variation in reciprocity, solidarity, self-determination, and resistance', *Environmental Justice*, 15(3): 160–9. http://doi.org/10.1089/env.2021.0046

Hess, C. and Ostrom, E. (2005) 'A framework for analyzing the knowledge commons', in C. Hess and E. Ostrom (eds) *Understanding Knowledge as a Commons: From Theory to Practice*, Cambridge, MA: MIT Press.

Edwards, P.N. (2010) *A Vast Machine: Computer Models, Climate Data, and the Politics of Global Warming*, Cambridge, MA: MIT Press.

Escobar, A. (2018) *Designs for the Pluriverse*, Durham, NC: Duke University Press.

Fisher, J.B. and Fortmann, L. (2010) 'Governing the data commons: Policy, practice, and the advancement of science', *Information & Management*, 47(4): 237–45.

Lehuedé, S. (2022) 'When friction becomes the norm: Antagonism, discourse and planetary data turbulence', *New Media and Society*. https://doi.org/10.1177/14614448221108449

Little, M. (2021) 'Relational social policy: An invitation to co-create a new approach to local government', London: Ratio Research, [online] 15 July, available from: https://ratio.org.uk/writing/relational-social-policy-an-invitation-to-co-create-a-new-approach-to-local-government/

Loukissas, Y.A. (2019) *All Data are Local: Thinking Critically in a Data-Driven Society*, Cambridge, MA: MIT Press.

Mattern, S. (2021) *A City Is Not a Computer: Other Urban Intelligences*, Princeton, NJ: Princeton University Press.

Milan, S. and van der Velden, L. (2016) 'The alternative epistemologies of data activism', *Digital Culture & Society*, 2(2): 57–74.

Møller, N.H., Shklovski, I. and Hildebrandt, T. (2020) 'Shifting concepts of value: Designing algorithmic decision-support systems for public services', in *Proceedings of the 11th Nordic Conference on Human–Computer Interaction: Shaping Experiences, Shaping Society*, New York: Association for Computing Machinery, pp 1–12.

Powell, A. (2018a) 'Alison Powell on data walking', *TMG Journal for Media History*, 21(2): 146–50.

Powell, A. (2018b) 'The data walkshop and radical bottom–up data knowledge', in H. Knox and D. Nafus (eds) *Ethnography for a Data-Saturated World*, Manchester: Manchester University Press, pp 212–32.

Powell, A.B. (2021) *Undoing Optimization: Civic Action in Smart Cities*, New Haven, CT: Yale University Press.

Powell, A.B., Ustek-Spilda, F., Lehuedé, S. and Shklovski, I. (2022) 'Addressing ethical gaps in 'technology for good': Foregrounding care and capabilities', *Big Data & Society*, 9(2). https://doi.org/10.1177/20539517221113774

Prainsack, B. and Buyx, A. (2017) *Solidarity in Biomedicine and Beyond*, Cambridge: Cambridge University Press.

Rosenberg, D. (2013) 'Data before the fact', in L. Gitelman (ed) *'Raw Data' Is an Oxymoron*, Cambridge, MA: MIT Press, pp 15–40.

Siemens (2020) 'Siemens presents new software for locating system', available from: https://press.siemens.com/global/en/pressrelease/siemens-presents-new-software-locating-system

Suchman, L.A. (2007) *Human-Machine Reconfigurations: Plans and Situated Actions*, Cambridge: Cambridge University Press.

Tsing, A.L. (2005) *Friction: An Ethnography of Global Connection*, Princeton, NJ: Princeton University Press.

Walker, K. (2019) 'An external advisory council to advance the development of AI', available from: https://blog.google/technology/ai/external-advisory-council-help-advance-responsible-development-ai/

The Contingencies of Urban Data: Between the Interoperable and Inoperable

AbdouMaliq Simone

Introduction: What is available as data?

Data, data everywhere. Never a moment's rest. Never an aspect of life not potentially convertible into indicating something besides itself, never unable to participate in a gathering of factors whose particular compositions indicate future behavioral dispositions or scenarios. Data reworks the fundamental ontological status of things, as they no longer exist for themselves or for their actual and potential uses for others, but rather as placeholders, momentary points of reference for an assemblage of futurity always in the making. In other words, things are basins of attraction – to use cybernetic vernacular – that contribute to the singularity of specific events, personalities, and operations: a contributing factor to why events transpired the way they did and what their likely implications are to be. A person does not like a particular item, object, or experience for itself without the possibility of that preference being converted into an indicator of some trend.

Data is the materialization of the potentialities of the transitive to exceed the prevailing grammatical rules that govern semiotic-syntactic relations. It is the form of existence – digital units of information – through which any entity can become transactional, comparable, or synthetic. Where the elements contributing to the operations of any situation, event, institution are analysable in terms of the proportionality of these contributions. As such, data has no meaning outside the architectures and operations of relationality, interaction. Data assumes the presence of a network, and a network is not discernible or operational without data, without entities functioning as positionalities or nodes in interaction. Data is not a discrete object as much as a mode of existence to be enfolded into a decision, legitimation, or prediction (Hui, 2016). In other words, data is a moment of discreteness, something that points to a state of affairs, event, or characteristic that will serve to steer future action. It is the product of a way of paying attention, to regard something as being appropriate to applications beyond itself.

As such, a politics of how data is created and applied in urban settings would seem, in my view, to respond to a series of key questions:

What do you need to know in order to reside viably in settings and circumstances where there are insufficient consistencies in how things operate – from authority, infrastructure, livelihood?

How do you style everyday performances – of self, household – so as to instigate displays of important operations for which there are no stable concepts, guidelines, or modes of appearance?

What is data in circumstances lacking the systematic and distributable organization of information – such as circumstances where important matters remain either uncounted, unspoken, or uncodified?

And when computable, digestible data does exist, or where various apparatuses of surveillance and monitoring turn everyday social action into data, beyond the multiplicity of dissembling practices that attempt to deflect scrutiny, how do various collectives manage their own need to know and exert some control over how they are known?

These are questions that are salient to residents across many urban contexts; where there may indeed be 'rules of the game', structures of authority, and patterning of distribution and opportunity, but which largely remain tacit, indiscernible, or the privilege of the few. For urban contexts are increasingly characterized by multiple and intensive uncertainties – derived either from an excess of disjunctions or intersections of seemingly contradictory forces or intentionally cultivated as a form of rule. Uncertainties are also instrumentalized as a medium of remaking, of shifting and recalibrating arrangements of all kinds. Again, as Hui (2016) emphasizes, data is not just one thing; there are not strict compositional rules about what data is or not. Rather, it entails the particular use of events, materials, and entities within the acts of making decisions, of trying to prove or justify particular courses of action, or to predict what is likely to take place as a result of specific actions. Of course, there is the data of repetition, of breaking activities and spaces into modular blocks that are associated analogically – described, compared, attributed as the purview of particular kinds of actors and activities. There are rituals of transmission, such as stories, instructions, and liminal conversions, that provide measures of assuredness, conviction.

But in many instances, things have to be incessantly 'figured out' phenomena provided a 'figure' that can be sensed and conveyed. The simplest account of data capacities would be through networks – familial, ethnic, associative, territorial, sectoral. Yet how do such networks produce and convey information, turn it into knowledge, valorize its enactment? In part – and only a part of the story – such an account needs to take place through orientations, perhaps an *operational ethos*. In other words, what kinds of data are available to whom, and given this availability, what is it that particular kinds of actors can thus do as a result. If anything that exists might potentially be construed or used as data, what kinds of things become operational as data and under what circumstances? How is the specific meaning of data negotiated, and how might particular materials or events that exist in the world be used, made available, that have no apparent use or meaning: that otherwise are *inoperable?*

Given this notion of the inoperable, the generation and use of urban knowledge thus requires forms of apprehension and conveyance that go beyond conventional modes of calculation, measurement, and value. What can be made available; how are people exposed to the world in multifaceted ways; how can they navigate urban contexts in ways that effectively make use of the ever-shifting relationships among things? These are also important considerations of how things are known and in what ways.

Within urban settings there is an increasing and simultaneous amplification of values and a dispensing of them. There is a need to define a specific place for oneself, an asset, something to be secured and precisely defined. On the other hand, in many cities, we witness a population largely on the move, foregoing the consolidation of place and destination in favour of itineraries of circulation through which people encounter various situations and other people with whom more provisional, temporary arrangements are forged. This duality of securitization and suspension in many ways is a mirroring of data ontologies. Algorithmic operations are directed toward specifications of value, of what counts most optimally among an increasing volume of possible relations. Yet, at the same time, any calculated disposition is always being reassessed recursively, generating continuous updates of the implications of various entities and variables interacting with each other. Put crudely, what makes sense in the morning, may no longer do so by the afternoon.

In Jakarta, for example, more and more people spend an inordinate amount of time on the move. Not just to get from one defined place to another, but also speculating about new opportunities and affordances; seeking the right 'position' at the 'right' time, knowing that any disposition of 'rightness' is only temporary. Households are spread out across the urban region to become nodes in a circulation of impressions, in a distributed agency of piecing together different opportunities, which are changing all of the time. Thus, what accounts for the decisions people make about how and where to move is often inexplicable to them, attributed often to 'maximizing' their exposure to the city, or having a 'feeling' about what might exist 'over there'.

Given such practices, urban justice is not the restoration of some overarching commonality, not the equilibration of difference through the fair apportionment of specific resources or opportunities. Rather, it is anchored in the availability of differences to generate new dispositions for living, without judgments of their efficacy in terms for which these differences – whatever their form – have not contributed to developing. This space of availability has been a critical historical condition of the possibility for 'Southern' urbanities to elaborate a capacity to sustain growing populations and differences. This capacity to enact different ways of doing things outside the debilitating capture by purportedly 'definitive judgements' of efficacy facilitated the formation of expansive associations. All of the different ways people, things, and places were associated with each other, with no guarantees that they would work, nor any expectations that they needed to be prolonged or institutionalized.

In Lagos, the long-honed street-level circulations of information (*radio trottoir*) were recently deployed through social media as an instrument to call attention to and disrupt the arbitrary and ruthless practices of policing and security, eliding the state's capacity to maintain the invisibility of its more vicious approaches to rule. The systematic and careful availing of the realities of local lives to a larger surrounds through practices of street level *testifying,* one neighbourhood to another across Freetown, enabled Leoneans to piece together an effective response to curtailing the impact of Ebola in the city. The circulation of TikTok music videos, music clips, WhatsApp videos enabled a sense of connection experienced across popular neighbourhoods in Luanda, where the state attempts to rule through fragmenting and dividing the city.

These are all works in progress but emphasize an ethos of availability, putting messages, images, experiences, and thoughts out in the world to be tried out, reworked, and fed back through elongated circuits of reciprocity. Women's associations in Agadez, a key nodal point in transmigration from across West Africa to the Mediterranean and increasingly a locus of European interdiction and surveillance, adopt passing migrants as their own children so as to attenuate conflicts about who belongs and who does not. As young people in Abidjan, for example, 'find' themselves across a wide range of 'versions' – jobs, hustles, identifications, locations – they depend on the consumption and display of particular brands or styles as a way to assert a specific and strong identity that serves as an instrument of continuity as they constantly change social performances.

These practices are more than simple resilience. They are forms of *computation* that try to recombine the materials at hand to find new spaces within which people can operate and make some kind of living. Such recombination is premised on the availability of materials to such new arrangements. At the same time, we know that availability is also thoroughly entwined with the implicit demands of multiple and contradictory requests, loyalties, responsibilities, and sometimes dizzying turns of events. It is difficult within a field of extensive circulations, expressions of experiences and demands, to make clear priorities as to particular courses of action, to always know what it is important to pay attention, or to work out a sense of proportion as to what kinds of events and actors are exerting specific forces over one's life.

As an inverse of atrophy, the risk here is simply overstimulation and the inability to discern and predict. Or investing in forms of narration easily digestible, curated for their affective 'punch.' Time might be wasted in seemingly interminable verification – trying to determine whether something is true or not. Anxieties may abound concerning possible contamination or possession: the worry that one is always available to be 'messed with' (Benjamin, 2019).

Availability can also be dissimulated: where things pose as something purportedly familiar, generic – but rather use the form of the generic to compress many different elements and for which is difficult to detect proportionality about just what kinds of factors are at play. For example, when we look at a large, seemingly faceless housing complex, one of thousands we may have already encountered, we would seem to immediately know what it is like, what takes place there. But sometimes things that look the same hold within them very different capacities and compositional elements. Such 'deception' actively unmakes the possibility of detection being the method through which individuals and populations are subsumed into a system of proportionality – whether they are more or less healthy, more or less immune, more or less eligible, more or less valuable. Instead, the generic here connotes a space or composition capable of holding within it things and processes that may be related to each other, or not – where what something is maybe multiple but does not owe its existence to how it is positioned within a network of multiplicities and through which it accorded particular statuses and potentialities.

Of course, such problematics, centered on the implications of unstable certainties, differential access to data and decision-making opportunities, and the possibility of many potential trajectories emanating from the same data sets, become an occasion for governance. Governance that attempts to regulate the implications of exposure to data; that is, who gets exposed, to what extent, and under what circumstances.

Governing data everywhere

Governance is said to be increasingly data-driven, led by empirics rather than ideological orientation or political expediency. What people actually do, rather than the ways in which they represent themselves or configure narrative explanations of their actions and aspirations, are to constitute the basis of decision-making. Narratives about how things got to be the way they are, in terms of genealogy or conjunctural analysis, are displaced by the emphasis on interoperability (Mackenzie, 2016). Here, the constitution of a unit, a piece of information, is designed to facilitate its translatability across a potentially infinite series of comparisons. It is the design of a statistical framework that enables information about divergent aspects of life to be correlated in such a way as to assess mutual causation and influence. Health, education, financial, medical, and social records about an individual, household, and institution are shaped in such a way as to render them comparable, and thus able to generate analyses as to their interrelationships and relative proportionalities of influence.

Importantly, the raw material of data, its existence as an entity to be enfolded into various assemblages and sites of calculation, is produced by the person or institution themselves. It is viewed not as an abstraction of experience but rather the conveyance of experience itself as material to be analysed through its comparisons with others. Existence is thus data, and data increasingly becomes the mode of valorized existence – to be scrutinized and affirmed by virtue of its interconnectivities and participation in generalizations, influences, and predictions.

Take Ruth Wilson Gilmore's conception of the anti-state state – an apparatus captured by those who don't believe in the apparatus as a means of curating a social body forged through equanimity and justice, and where the demands of social reproduction – devolved to scales and institutions insufficiently equipped or willing to provide for basic needs – are partialized: subjected to both incompletion and preference, of increasingly intricate accounting systems of whose life counts and in what ways, and in what can be legitimately extracted from them. Here governance becomes that of a corporate accounting system, a continuation of the double accounting system that was an integral element of the economy of transatlantic slavery. It becomes a continuous account of which lives count for what kinds of affordances, and what can be legitimately extracted from them (Gilmore, 2022). In the constitution of a body politic, then, data is everywhere.

If data is everywhere, it is not just its ubiquity that is important here, but the ways in which it cultivates a sense of *everywhereness*. Here any singularity becomes potentially salient anywhere, can be mobilized as a referent to make an assessment about the functioning of anything else. While things remain anchored to their niches and ecologies, they are also simultaneously detached from them, capable of showing up as a potentially meaningful variable anywhere. Even the most seemingly marginal of places or occurrences can be mobilized to have something to say about events and phenomena far removed from their apparent ambit. But the real importance of *everywhereness* rests in vastly expanding the domain of exchange, where a means is worked out to render the discordant potentially translatable and convertible; where nothing is off limits in terms of its availability to add or subtract value (Stiegler, 2018).

If data is then everywhere, we are also all exposed to it. What do we do with such exposure, and what kinds of exposure provide specific kinds of affordances? The engagement and use of knowledge and data goes beyond simply practices of acquisition, to encompass ways of being exposed to the world, to what is 'out there', often without specific definition or location.

On the one hand, urban residents are exposed to greater levels of scrutiny, which may entail enmity, conspiracy, cynical judgments, dismissals, and indifference. Exposures to vastly elaborated extractions of what bodies can do under varying circumstances: how can they be manipulated, enrolled, lured, held, and indebted. As such, immediate circumstances are often ridden with petty antagonisms, undone expectations, and disregard. Yet exposure is frequently referred to now self-consciously as an exigency, a need to be exposed, and to something that is often nebulous, uncertain (see Harcourt, 2015).

Notions of exposure also inform the logics of the construction and use of built environments. Operations of 'infrastructuring' go beyond a genealogical scope – go beyond how things got to be the way they are – to inscribe actors in an ongoing series of interactions, forms of witnessing and gathering, and modes of 'being together.' Such infrastructuring creates a particular kind of exposure to the larger world. An exposure that ensconces actors in materialized sensibilities of encounter that are specific to the immediate environs in which they operate. This is something not easily 'messed' with because it 'refuses' being definitively known or parcelled.

On the one hand, the borders between urban territories constantly shift between administrative designations, zones of social intimacy and emotional attachments, circuits of everyday mobility, and shifting forms of authority. Yet, the intersections among conduits of circulation, spaces of relative domesticity, the modulations of public and private interaction, the routines of everyday social reproduction, and the vectors of sensation marked out by the materials and designs of built forms generate a specific orientation and capacity, a *specific* imprint on the larger urban surrounds. This is neither the only orientation nor the only impact. But it is something specific, immeasurable, and untranslatable, that infrastructuring makes possible.

Those who undertake new lives just beyond the city are exposed to situations where there may be little in terms of institutional support or urban services. There is the exposure that comes with concluding that one's familiar ways of managing daily affairs are insufficient, and that the present composition of family and friends may not be enough to keep up with things. Here is exposure to new circumstances over which one cannot exert much control. The itineraries of those residents trying to engage new opportunities, new incomes, and new places of everyday operation result in their exposure to multiple contestations, power dynamics, and forms of authority with which they have only limited understanding or ways of dealing. All around are those prepared to take advantage of their vulnerabilities, their desire for some sense of direction. But they also see others around them take inordinate risks to do something different with their lives, and sometimes they also see the evidence that these risks indeed work. Whatever the game, they refuse to be construed as victims of it.

Exposure implies that there is always something to be done, some project or projection of self, which simply in its announcement acquires a certain efficacy. Exposure is not simply a revealing, or the act of informing and being informed, for it brings into existence not only a situation but a conundrum, even a challenge. For if something else is now being made known, then the context or reality to which we have adapted to up until now is also revealed as insufficient.

But there is also an impression created that there is a market anywhere and everywhere for what could be offered. As such, the sense of contextual relations is often set aside. Things don't emerge from the characteristics of the interactions among specific contextual features but more in terms of fiat, exposure (the curation of an Instagram page), and the design of a message to lure likes and followers. The profusion of stylized images produced by the masses across even the most seemingly marginal or problematic of cities generates a reality beyond the conventional demographic, social, or economic markers (Degen and Rose, 2022). While the cultivation of fashions, appearances, and tagging may remain anchored within specific cultural sensibilities, there is an impression that what is being presented could be taking place anywhere and everywhere.

My own Instagram account, tailored to the streets of Sahelian cities, daily demonstrates the capacity of participants to respond to a 'world' that is not proportionately skewed to any single place. For participants are less interested in representing or embodying a particular cultural identity than in generating their ideas about what being part of a world never fully in existence might comprise of.

So, data implies use and users – decisions, anticipations, and scenario-building. The user is not a predetermined subject who remains unaffected by the use of data. Rather, the user becomes a subject position always 'thrown off', destabilized by the data attained through processes of calculation that have become increasingly recursive and self-referential. Rather than processing data through algorithmic operations that are traceable every step of the way, those processes are incessantly recalibrated through their own histories of operation that interrelate data according to their 'own experiences.' For information has its own indeterminate characteristics that are actualized only in the algorithmic procedures themselves. This recursivity is the presumption of machine learning and general adversarial networks. As Yuk Hui points out, users are already 'part of an algorithm' that is not only 'part of a database' but is also, in part, constitutive of an algorithm's 'executability' (2018, 32).

In order for computation to do things, it already has in 'mind' the positions of the users, not as subjects replete with intention and needs, but more as instigators, part of the overall environment in which these processes are situated. The stability of the user is precisely that which is to be questioned, that which is only a provisional and insufficient arrangement in an overall situation of disequilibrium, which data navigates not to bring order to a situation but to maximize the potentials of instability, freeing any experience or object to participate in new configurations of sense-making (Hayles, 2021). This is why specifications of the character, function, and responsibilities of citizenship seem to make little sense in contemporary configurations. Particularly as urban economies seem increasingly driven by maximizing the process of urbanization itself – urbanizing, in other words, complexifying and multiplying the kinds of 'collectives' through which individuals might be identified, gathered, even if such collectives are entirely virtual.

If computation concerns relationality, then the relations operationalized in calculation always potentiate an undermining of the stability of those things that are brought into relation. Take Ramon Amaro's (2021) explorations of the role of blackness as data. If blackness is operationalized as an instrument for securing the stability of particular kinds of individuations modelled on property, and where freedom is premised on the capacity to develop property as one sees fit, then what happens when it is incumbent on blackness to endure in this relationship without recourse to the trope of individual subjectivity, when for all practical purposes the black individual is not a subject? An entire domain of operations is left open, not within the proper place of the human, but where the (non)human body is able to extend themselves into articulations with non-propertied relations beyond the framework of recognition and value. Such possibilities continue to introduce new instabilities into the racialized system of social ordering, which are in turn continuously misrecognized as that which requires an intensification of racial control. As Amaro indicates, the black individual is always already preformed as that which sustains the illusory characteristics of whiteness, not so much as a subjugated body lacking the capacity to be a fully fledged subject, but rather as a body outside of subjectivity and without which the position of the free white individual would be impossible.

For individuation – the crafting of some balanced synthesis of forces – is the provisional working out of new problematics generated by entities acting in the world. The individual is an embodiment of tensions and temporary resolutions, whose actualizations generate information that any particular stage of individuation cannot fully use or contain, and which provide possibilities of disjunction. As conditions and responses are remodulated, the crystallization of value – such as the valuation of epistemic certainty as the way of orienting oneself in the world – becomes transduced into other arrangements. This takes place through a series of translations in which value acts as the consistent marker – such as the presumption that the white individual is the determiner of his fate. Yet it is a sovereign individual who can only perform that sovereignty through its curtailments in others. It exercises its prerogative as a self-reflecting entity with an interior life by defining an 'other' whose entire being is read on the surface of the skin, even as their endurance entails a vast reworking of flesh beyond individual form.

If urban governance, then, increasingly relies on continuous recalibrations of what bodies, selves, events, and materials are according to shifting computations and ways of putting things in relation to each other, then how is it possible to navigate urban spaces and conditions? If social contracts and forms of consensus seem increasingly devalued, what kinds of stability are possible in such data fluidity? How can one negotiate with urban conditions that often seem beyond negotiability? But first, in order to address these questions, what kind of urban atmospheres are we 'up against', and how does the urban act as a data environment?

Data and the production of uncertain urbanities

The design of buildings, projects, budgets, capital investment plans, strategic visions, intercalibrated infrastructures of material flow – water, power, waste, information, transport – are increasingly generated through constantly updated relationships among increased volumes and types of data. As such, a space of deep and extensive relationships is being created that impacts the 'real city' but which is something else 'besides' it even as it is a 'part' of it – something perhaps akin to 'data doubles.' Here, the interfaces are uncertain even as the pragmatics of these calculations emphasizes the sense of stability and order brought to bear on the 'real city'.

The MIT Sense Lab, for example, using LiDAR (3D laser) technology can trace a million data points a second to provide a morphological representation of informal settlements, making visible spatial organizations that otherwise would remain beyond visualization and, thus, systematic intervention. Whether or not the residents of Rocinha (Rio de Janeiro), one of the settlements mapped, can engage such representations and make them useful to their own understandings, aspirations, and plans remains a question. As does the capacities and interest of municipal administrations, particularly as territories such as these are affectively complex atmospheres overlaid and experienced with an expansive repertoire of sensations and hypotheses. Rather than such representations constituting an expressive abstraction of 'real urbanization processes', they constitute a mode of urbanization themselves, setting form, materials, functions, and actors in unprecedented relations with one another, positing new modes of existence for any component of the territory. These modes can intersect, run parallel, or abnegate each other depending on who invokes and uses them.

While the use of such mapping technologies might be motivated by the urgency to draw on and maximize an enlarged repertoire of urbanistic sensibilities and resources, a process of valuation is expanded that potentially makes judgments about whose lives count and in what way. Who is sufficiently resourceful or resilient? Whose bodies can be moved around and tested? Given the increasing volume of terrain that is deemed uninhabitable – in terms of the by-products of climate change, security, or toxicity – decisions will be made about the terms of plausible interventions. Will people in certain circumstances be even more relegated to their devices, or become the objects of systematic displacement, relocation, or even tacit culling in terms of investing in those most likely to either survive, flourish under uncertain conditions, or comply to existing norms or political quiescence?

Entire infrastructures of data collection and production are distributed across urban landscapes. Sensors, processors, and actuators are designed to intersect software, hardware, action, behaviour, and movement, so that trajectories of force and impact can be detected and registered. These forces can be mechanical, thermal, biological, chemical, optical, or acoustical, to name a few. Force, impact, and movement then are registered according to parameters.

Parameters track the composites of action, the ways in which different circulations, behaviours, and operations affect each other. This is not just about the operations of 'smart buildings', where the power generation, atmosphere controls, lighting, wired and wireless infrastructure speeds and capacities are automatically calibrated to the oscillating uses or 'statuses' of the building at any particular time. The range of modulated impacts to be registered can also be built into large-scale tracking systems. In this way, for example, patterns of rainfall, traffic flow, water usage, power generation, capital budget allocations, density levels, and service consumption can all be brought into some kind of relationship with one another, where their impact on the operations of one another can be subject to measurement, the production of quantitative values.

How, then, does the body move through their exposure to these various relationalities, these ways of calibrating everything in terms of their actual or potential relationships to everything else? How all-encompassing are these data systems in terms of how we actually navigate everyday urban life? The entire environs is sensate, in that all materiality pays attention to what it 'touches', and every material, and thus project, is a conduit of forces and histories passing through. Whatever exists in place is a product of things being carried and displaced, and every place is a node from which lines of flight ensue. The matters of where to go, what to do, how to do it, while embodied in a series of routines and culturally sanctioned practices, are not only being incessantly, even if only slightly, revised in terms of responding to new contingencies but also retailored to provide options and alternatives when familiar conduits begin to decay or need to be repopulated by new personnel or techniques.

It is important to emphasize that data politics is not just an 'us' versus 'them' divide; not a matter of an all-encompassing surveillant state – even if China now presents a model of such. The experience of *everywhereness* talked about earlier also refers to the domain or scale of operations. Seemingly everywhere people are using the possitiblities availed by data and its mode of existence as a means of exerting a force, either in terms of making something known, making something happen within a particular interest or speculation, or creating a version of reality.

For example, the Betawi of Jakarta – the city's hybrid 'original' inhabitants – long sought their value in land rather than in wage labour. They long resisted incorporation as a salariat, but a resistance that left them vulnerable to increasing accumulation through urban development. Many were compelled to sell their land and move to the periphery. After years of progressive economic marginality, witnessing the skyrocketing rise in value of their former landholdings, and faced with intensive land-grabbing even now at their peripheral locations, many Betawi are trying to respond by exercising a particular kind of political force.

Through their active participation in various Islamic movements aimed at integrating sharia into all aspects of social life, roving convocations are held across hundreds of mosques in the region. This is important because no matter the development trajectories of particular neighbourhoods and districts, mosques cannot be removed; they legally must remain situated in their original territories no matter how massive the development. These convocations are used as a platform to gather and consolidate data on the district, which is compiled and circulates through specially developed 'religious' apps that detail land transactions and compile an inventory of home-based and small enterprises, providing a circuit of commodity exchange.

The apps are also a surveillance tool in terms of watching the actions of local authorities, power brokers, and municipal agency personnel, spreading not only information about their functions and behaviours, but spreading rumours, fabricated histories, and events. Workshops that fill two single streets in the nearby suburb of Bekasi turn out hundreds of Twitter accounts and Instagram pages that not only weigh in nearly everything, but also 'create' events and transactions that exist only in this medium. If tangible landholdings have been lost, and thus the basis of Betawi wealth and influence, then this conflation of an Islamic movement with network building, data piracy, social media, and low-level commerce becomes a means of Betawi taking back space, creating a form of virtual land from which to exert political influence.

In today's urban life in general, given the production of ambiguity, ambivalence, dissimulation, and transparency in varying rhythms, and sometimes all at once, continuous questions are raised for residents about what to show and how, what to make available and in what degrees of exposure. Our understanding of commonality is usually and tacitly affirmed through a shared series of problematics and everyday exigencies. The information required for styling everyday performance comes through social exchanges that continuously hold something back in order to elicit the inquisitiveness and inquiries of others, which are taken as demonstrations of intent. But what happens when such social exchanges become more indiscriminate, arbitrary, and conducted across multiple locales? What becomes of navigation then?

Arrangements of built environments are not only manifestations of property relations, policies, and affordability but are materializations of navigational circuits, of all the concrete histories of residents making themselves available in specific ways to one another. Perhaps more importantly, navigation is driven through an active unknowing or social distancing from established familiarities and assumptions. A distancing from any clear bifurcation of state, citizen, and subject.

It is an act of deferring any definitive 'settling in' of a dominant mode of valuation or ways of attributing significance and authority. Hedges against debilitating uncertainty have to be balanced with the warding off of entropy that comes from overly static orientations. After all, the density of materials, social compositions, livelihoods, and performances always suggests a wide range of possible inclinations and ways of doing things whose curtailment requires excessive expenditures of effort. The costs and advantages of holding things down must constantly be weighted with those of letting things go, even if people face operating completely in the dark, or at least feeling as if they do so.

Navigation also includes a process of *intermediation*. As urban regions are multiply exposed to a wider range of financial, logistical, and cultural flows at a global scale and an intensified particularization of individual sentiment and practice from below, any built environments, spatial arrangement, and economic and administrative function has to mediate and provisionally resolve a multiplicity of often contradictory or, at least, not easily synthesizable considerations. As urban regulatory environments are replete with exceptions, loopholes, and temporal qualifications, *popular economies* are aimed at straddling the divides between their apparent compliance with or subsumption within normative frameworks of operation, yet at the same time stand aside, reserve something not quite 'on the books.'

For example, the vast networks of textile production in Jakarta on the one hand reflect the ways in which large-scale factory production has been decentralized into hundreds of small units, each working on specific facets of clothing production –cutting, patterning, sewing, stitching, buttoning, designing – all vertically integrated into a few large corporate structures. But there is also a substantial lateral chain of production and marketing from these very same units that has been developed over time through intersecting memberships in religious and women's associations, impromptu popular markets, unions in the ports, and eating places where truckers congregate.

All in their own way have paid attention to how various loopholes, tenancies, friendship networks, land statuses, and ethnicities could be capitalized on in order to not only supplement the incomes of the players involved but to create an infrastructure of transactions that can be mobilized for influencing political authorities, investing in affordable housing, and improving urban services. Whatever gets standardized or formalized as normative relations and the methods for calculating them; whatever exists as 'data-driven' decisions also points to possibilities beyond the dispositions they would seem to potentiate or enforce. No matter how precise the process of generating probabilities and best solutions might be, the very process also points to a world of contingent relations, relations unanticipated before and during the calculations applied.

Conclusion: The unforeseen potentials of data production

The computation of relations does not exclusively coincide with the reduction of temporal qualities to preset probabilities. Instead, it reveals the formation of another space-time and describes the simultaneity of experiences without reducing distinct spaces to the relativity of lived time. For example, built environments live simultaneously within multiple temporalities, such as the time of material decay, of particular kinds of uses, the time of metabolic operations, to name a few. The space between things is always full of 'things,' visible or invisible under varying circumstances and technical applications. This continuity of things as they form, deform, and reform, through the processing of invariant functions, is continuously 'interrupted' by 'events.' Unanticipated occurrences. Parameters thus 'meet head on' in ways that are not easily 'figured out.'

Such logic can be extrapolated to consider the relationalities of various components of the urban built environment, regardless of the degree to which they are equipped with sensors. The remnants of old construction – residences, workshops, sheds, that were situated along circuitous narrow lanes, dead ends, switchbacks so as to both avoid and accommodate different claims and interests – meet head-on with the vestiges of public parks never used but which bear the name of national heroes whose memory could never be affronted. These meet head-on with the intricate constructions of dwellings whose unfinished upper stories are intertwined across pylons and wires and planks that act as alternative thoroughfares to those at street level. These meet head-on with the massive vacancies of parastatal landholdings long intended for every conceivable development project but in the end simply make-up for interminable budget deficits. And these meet head-on with tightly drawn and dense quarters that now abut major commercial zones and hurriedly add on whatever rooms they can to available living quarters in order to accommodate low-wage service workers. All of these registrations of force may never be the objects of parametric measurement or reasoning. But they metaphorically indicate the problems with parameters in terms of their being designed to focus so exclusively on preset entities and their functioning.

As more elements are set in relationship to one another, a larger plurality of eventualities is created, as well as milieux populated by 'strange creatures' – hybrid entities, new forms of transmission and conveyance that are not searchable or are always changing course. For the grounds on which digital and algorithmic governance endeavour to guarantee specific decisions are themselves predicated on operations which by their nature have no guaranteed course of action. While algorithmic operations attempt to affect what specific entities will do, they themselves need make no reference to a physical world. They bypass any form of representation, potentially discovering actualities for which we have no language. What seem to be excessive efforts to precisely calculate the urban future in the end, perhaps, make it more uncertain. Despite calculations that render distinct parameters interoperable, parameters to be interrelated 'bring with them' data of the past, entropic information, and histories of being facets of other relationships, and thus point to incalculable potentialities, which cannot be incorporated into an overarching system of continuous variation or resilience (Parisi, 2013)

The question is, who is able to take advantage of such contingencies – that is, the prospect that completely unfathomable realities could emerge through the growing infrastructures of computation, interoperability, and parametric design?

Are these simply matters for the rarefied domains of physicists, engineers, and mathematicians working in collaboration with hi-tech corporations and militaries? How visible and public are the massive farming and analysis of information anyway? For professionals narrowly and rigorously trained in niche settings and points of view, how open are they to events of surprise, where a true appreciation for what they confront means being dislocated from almost everything they might know? How might we move from the ways in which technical apparatuses deploy contingency to more popular, everyday uses by 'ordinary' people?

As Ravi Sundaram (2015) has pointed out, data production, retrieval, and deployment is subject to a wide range of assumptions about what constitutes the truth of any situation, as subalterns attempt to upend the hierarchies of knowledge production and verification with their own data practices.

> In the contemporary digital era, this is a neurophysiological zone amplified by the mix of mobile computing objects, moods, and sensations. Provisional networks form around these temporary connections: Bluetooth sharing of media by sailors, urban proletarians, and migrants; shadow libraries moving via USB drives; hawala transfers via text; neighborhood shops that refill phone memory cards with pirate media. Online shadows exist in WhatsApp sharing networks, dancing around regimes and mobile company filters. This is a remarkable infrastructure of agility and possibility.
>
> (Sundaram, 2015, 9)

Perhaps even more significant are the ways in which urban relationalities of data – that is, the process of constructing interoperability among diverse strands of urban life – is an analogue of working-class local governance practices across many districts of the world. While perhaps remaining at the level of a homology rather than a direct modelling, nevertheless it demonstrates the ways in which the specific composition of local social stabilities draws on what Simondon (2016) has called *the transindividual*. For the very concept of the interoperable requires a process of actualities exceeding themselves, drawing from a vast repertoire of possible actions materialized in its very interactions with the larger world.

On many occasions I have talked about *urban majority* districts – those comprised of an amalgam of the working poor and lower middle classes. As such districts have long demonstrated, a density of intersections produces multiple perspectives, where everything that exists is being recalibrated, repositioned in their relationships with one another because they are constantly being worked out and engaged by people and materials who are themselves continuously similar and different in different ways by virtue of these interactions. Density means not just packing in a lot of things into a limited space. Rather, it is the creation of a particular kind of space where people with their devices, resources, tools, imaginations, and techniques are always acting on one another, pushing and pulling, folding in and leaving out, making use of whatever others are doing, paying attention to all that is going on, fighting and collaborating.

Here, the example I frequently use is the Kebayoran Lama nightly produce market in Jakarta. The market may appear to be a gathering-up of individualized entities, albeit with different scalar composition. It may appear to simply be a place and occasion for buying and selling based on price advantages, sourcing of products, and individualized calculations of both the pressures of supply and demand and other larger price-setting mechanisms. But behind these appearances is an intricate infrastructure of understandings and data practices that enable these appearances to operate in concert.

Those who unload, deliver, park, invoice, sell, clean, buy, repair, instruct, smooth over, enforce, inform, circulate, allocate, juxtapose – all essential practices in the market – may be distributed across specific roles and individuals, but these roles, for the most part, can also be assumed by anyone operating within the market, and this way practices interoperate one another, as do those that perform them. In other words, the market involves actors that pay a great deal of attention to how all of the non-coherent practices work or do not work with one another, the oscillations of transactions and performances, the effects that a wide range of actions and behaviors seem to exert on one another. All of these intermeshings and frictions elaborate various story lines. Particularly for the brokers and local 'officials' who govern the market, there is a need to be the avid collectors of stories. They listen to various reports, observe the wheeling and dealing of the assembled characters engaged with the various affordances, infrastructure, and routines of the market. The power of the market largely is concretized in unsettling the dominance of any one story, a story that might break the ongoing line, an ongoing computational process. As a result, what we might come to know about this dimension of urban life is always already shifting, being repeatedly created anew, and also in significant ways enduring, stabilizing.

How might we then relate these more quotidian vernaculars of computation back to a different perspective of computation in general? How do the contingencies of everyday datafication and decision-making potentially correspond to a more open-ended, contingent domain of calculation? How to, in Sylvia Wynter's terms, develop a new code, a new narrative that specifies an ontology of the human beyond the genre of whiteness, as an integral aspect of the self-constituting operations of the human? Operations that have been historically attributed to the inevitability and naturalness of the modernist subject and its economic imperatives. Here the work of Luciana Parisi is particularly useful in demonstrating how contemporary processes of computation suggest an urban landscape or atmosphere that continuously *urbanizes* itself beyond what we recognize to be the city, suburbs, or periphery.

It is not clear what such urban formations are, but they are domains that break out of modelling the conventional definitions and parameters of urban space, the ways in which urban life is categorized according to specific forms, zones, functions, sectors, and populations. As 'smart' cities try to make the relationships among these categories both interoperable and generative of specific decisions and probability scenarios, they are subjected to becoming digital entities amenable to algorithmic procedures. But even here, as Parisi insists, the techniques of operating an interaction between the abstract materializations of these digital entities – the building blocks of calculation – posit unanticipated and indecipherable meanings and propositions that suggest a completely different way of being in the world.

Again, we tend to think of computation as a means of deciding what counts and in what way, which lives have value and which do not. Such determinations demand an incessant particularization of the world, breaking it down into differences that are weighted in terms of each other. Differences that exist only to be compared according to an unyielding equation, such as man = human = white. Differences that were compared computationally as a way of both reaffirming and extending the terrain through which this equation could be identified and applied. But as Parisi (2021) points out, 'indeterminacies become enfolded in the interactions of localities. As such, indeterminacies do not simply demarcate how meaning can change through use but rather how *techno-semiosis brings forward alien meaning or know-hows of another language that does not seek to match symbolic inputs with outputs*' (p 5).

The interactions affected by computation affect the computational process itself, meaning that there is no direct correlation between inputs and outputs, even as there is an interdependency among them. The ways in which algorithms operate – the codes specifying what is to be interrelated under what conditions – and the symbolic logics of how entities are to be analysed, how they relate to each other, interact in ways that generate information that cannot be completely determined, as they 'spread out' across other interactive possibilities that algorithms are 'discovering' all of the time; that is, relations that exceed the program and its symbolic vocabularies.

Instead of the dominance of interoperable relations, we have the materialization of the *inoperable*. Cities contain layers of inoperable relations that give to the urban its unique aesthetics without themselves ever being actualized. They are the stories that the city tells to itself about itself. Stories that are sometimes glossed as the sacred dimensions of the urban, its poetical magic, its dissolution of discernible scales and zones, its seemingly chaotic profusions of cause and effect. Sometimes these stories only manifest as a whisper that cannot be heard by anyone except the narrator, for whom there is no difference between telling a story and listening to it. These urban narrations seem timeless precisely because they change at the same pace as the worlds of which they are part.

Inoperable relations reset the scale of the city, then, by ceaselessly changing the modulations of social life as if they were always already in sync with a pulsating aesthetics that manage to desist actualization. As the city's identical but constantly mutating anti-twin, the city's numerous clusters of inoperable relations constitute shadow economies of urban affordances that both human and non-human inhabitants are captured by but which cannot itself ever be captured. Thus in its mobilization to act precisely and decisively with complex urban conundrums, the urban as we know it becomes perhaps more amenable to analysis and control; but at the same time, the urban takes a big leap in going beyond all that we might know now.

References

Amaro, R. (2023) *The Black Technical Object: On Machine Learning and the Aspiration of Black Being*, Berlin and Cambridge, MA: Sternberg/MIT Press.

Benjamin, R. (2019) *Race After Technology: Abolitionist Tools for the New Jim Code*, Cambridge: Polity Books.

Degen, M.M. and Rose, G. (2022) *The New Urban Aesthetic: Digital Experiences of Urban Change*, London: Bloomsbury.

Gilmore, R.W. (2022) *Abolition Geography: Essays Toward Liberation*, London: Verso.

Harcourt, B.E. (2015) *Exposed: Desire and Disobedience in the Digital Age*, Cambridge, MA: Harvard University Press.

Hayles, N.K. (2021) 'Three species challenges: Toward a general ecology of cognitive assemblages', in S. Lindberg and H.-R. Roine (eds) *The Ethos of Digital Enviroments: Technology, Literary Theory and Philosophy*, New York: Routledge, pp 27–45.

Hui, Y. (2016) *On the Existence of Digital Objects*, Minneapolis MN: University of Minnesota Press.

Hui, Y. (2018) 'Preface: The time of execution', in H. Pritchard, E. Snodgrass and M. Tyżlik-Carver (eds) *DATA browser 06: Executing Practices*, London: Open Humanities Press, pp 25–34.

Mackenzie, A. (2016) 'Distributive numbers: A post-demographic perspective on probability', in J. Law and E. Ruppert (eds) *Modes of Knowing: Resources from the Baroque*, London: Mattering Press, pp 115–35.

Parisi, L. (2013) *Contagious Architecture: Computation, Aesthetics, and Space*. Cambridge, MA: MIT Press.

Parisi, L. (2021) 'Interactive computation and artificial epistemologies', *Theory, Culture & Society*, 38(7–8): 33–53.

Simondon, G. (2016) *On the Mode of Existence of Technical Objects*, translated by C. Malaspina and J. Rogove, Minneapolis, MN: Univocal.

Stiegler, B. (2018) *The Neganthropocene*, translated and edited by D. Ross, London: Open Humanities Press.

Sundaram, R. (2015) 'Post-postcolonial sensory infrastructure', *e-flux journal*, 64, [online] April, available from: http://worker01.e-flux.com/pdf/article_8997956.pdf

PART II

Strategies

6

Experiments in Practice: New Directions in Municipal Data Policy and Governance

Sarah Barns

Today's digital platforms, and the ecosystems of sensors, interfaces, and protocols that allow them to deliver responsive digital services to users, perform in ways that are producing radical asymmetries in data access and use, accelerating the divide between those who command vast amounts of training data, through the ownership and operation of essential digital services, and those who don't. If the 'digital divide' has previously referred to inequalities in levels of access and use of digital technologies, today, increasingly, we see these inequalities expressed through the terms of access and utilization of data flows enacted through distributed platform ecosystems. As platform ecosystems continue to operate 'beyond the state' (Swyngedouw, 2005) and beyond transparency in the algorithmic operations of digital services, critical questions remain about how governments should negotiate more equitable data governance conditions in their custodianship of shared urban spaces.

In this chapter I discuss how city-scale, municipal reform practices are setting new directions in data access and availability that respond to the dominance of platform services across cities. As I discuss, experimental governance models are being tested which set out novel conditions around how platform providers generate and distribute (data) value. As these experiments show, municipal actors and organizations are using the urban scale to test and coordinate novel responses to multiscalar platform ecosystems, pointing to a role for the custodians of urban spaces to contribute to wider policy and governance debates around the future of data politics and data policy in a platform society.

From little things big things grow: understanding platform ecosystems

Contemporary cities operate as spaces for a wide array of digital platform services which utilize data and technology to offer responsive, data-driven insights and tools that effectively coordinate the interactions and needs of diverse users. The urban manifestations of these platforms, considered through the lens of 'platform urbanism', extends interest in the spatial implications of digital platforms away from specific technologies, interfaces, and apps to the range of organizational and relational practices and protocols reconstituted through platforms today, which extend to include standardized approaches to data sharing policies, protocols, and business models (Helmond, 2015; Barns, 2020; Söderström and Mermet, 2020). I use the term 'platform ecosystems' to describe how digital platforms reconstitute existing urban relations and transactions through integrations of code, commerce, and corporeality. A focus on platform ecosystems is intended to highlight the multiple intersecting (digitized) relations that are coordinated through platform services, drawing attention to how platform companies intermediate and in turn reshape the wider socio-spatial conditions and relationships they operate within – often deliberately in ways that accelerate broader network interactions and, in turn, data accumulation (Barns, 2020, 100).

The platform-as-ecosystem perspective has been widely championed as a core strength of platforms as commercial disruptors, but it also remains critical to an emergent data politics of critical platform urbanism. In the early 2010s, platforms and their pundits sought to position themselves as enablers of 'open innovation' and reform, because unlike traditional firms, they allowed external actors to extend and improve their operations (Simon, 2011). 'Platform-based competition' (Tiwana, 2013) was thus seen as a shift from more vertically integrated companies to more porous 'platform ecosystems', where the utility of platform services were evidenced by the scale of the ecosystems that surrounded them. As one platform pundit described it: 'We are not in the business of building software, we are in the business of *enabling interactions*' (emphasis added) (Choudary, 2015). In a platform society, technology-based competition was thus seen to shift from 'a battle of devices' to a 'war of ecosystems' (Armstrong, 2006).

Widespread attention towards the range of social, economic, and political implications of platforms today recognizes the power of platform ecosystems in coordinating a range of multisided interactions and transactions. This coordination role is ultimately achieved through data governance protocols and modes of data exploitation instituted by commercially focused platform owners. Mackenzie (2018) has used the term 'proprietary opacity' to describe conditions of infinite programmability established within the ecosystems of platforms, by virtue of their APIs (application programming interfaces), which at once actively decentralize data production and use, while simultaneously recentralizing data collection and governance (Helmond, 2015). These relatively porous boundaries existing between a platform company and its wider ecosystem of users and agents have been central to the way platform companies have been able to discursively position themselves as disruptive, value-creating entities (Gillespie, 2010), allowing anyone with a smart phone or internet connection to get on board and join the 'sharing economy' via Airbnb, Airtasker, Uber, or other urban platforms (Figure 6.1). But if the 2010s saw platform companies and their ecosystems of diverse users, interfaces, and protocols marketed as pioneers of disruptive open innovation, the 2020s has seen much closer critical attention towards how platform ecosystems enclose their users through new data dependencies, data surveillance methods, and algorithmic 'nudging' of behaviours that prioritize and entrench platform-focused interactions (Langley and Leyshon, 2017; Raetzsch et al, 2019; Boeing et al, 2021; Bauriedl and Strüver, 2022; Sadowski, 2021).

No longer simply an app to facilitate ride sharing and 'disrupt' laggard taxi services, Uber Technologies, Inc., for example, today acts as a global data ecosystem, whose increasingly diverse applications across multimodal forms of transport aim to intermediate more and more diverse forms of human mobility in contemporary cities – to become the 'platform for human movement' (Fowler, 2015; Domurath, 2018; Barns, 2021b). Uber's ecosystem is not only its drivers and riders; it also includes open source licences, external proprietary platforms, data centres, cloud computing services, subsidiaries, and broader supply chains, whether of vehicle service providers or hotels, airports, and other industries that are integral to its capacity to deliver mobility services across the world. As it instruments this ecosystem into action via the Uber API, Uber institutes common protocols that ensure diverse data is collected, stored, and used in standardized ways (Domurath, 2018; Pollio, 2021; Barns, 2021b).

As platform ecosystems like Uber have expanded, greater asymmetries of data access have been accompanied by growing concerns over the lack of transparency in the design and applications of platform governance and pricing models, including algorithmic pricing and other data-intensive machine-learning (ML) tools. For example, a proprietary ML application called 'Aerosolve' is responsible for determining how Airbnb hosts should price their listings, and it is widely believed that prices set by the algorithm are artificially low, so as to maximize the appeal of the Airbnb platform to renters, thereby undermining the commercial attractiveness of competitors (Boeing et al, 2021). However, there is no way for external authorities to determine whether this is the case without access to the Aerosolve algorithm, which remains the commercial property of Airbnb. While a platform like Airbnb has been required by local regulators to provide more data from its

Figure 6.1: Platform ecosystems infographic

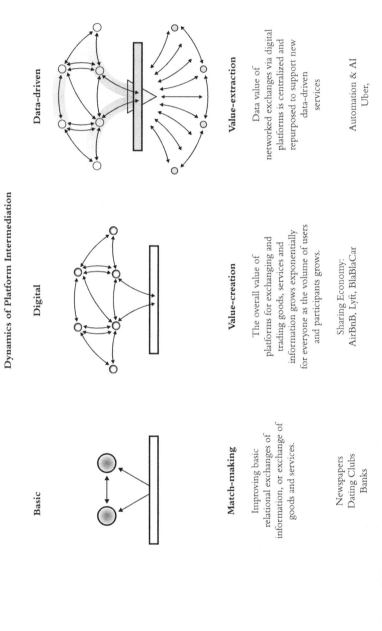

Dynamics of Platform Intermediation

Basic

Digital

Data-driven

Match-making

Improving basic relational exchanges of information, or exchange of goods and services.

Newspapers
Dating Clubs
Banks

Value-creation

The overall value of platforms for exchanging and trading goods, services and information grows exponentially for everyone as the volume of users and participants grows.

Sharing Economy:
AirBnB, Lyft, BlaBlaCar

Value-extraction

Data value of networked exchanges via digital platforms is centralized and repurposed to support new data-driven services

Automation & AI
Uber,

Source: Barns, S. Platform Urbanism: Negotiating Platform Ecosystems in Connected Cities (Palgrave Macmillan, 2020).

platform in order to facilitate improved regulatory oversight, algorithms such as Aerosolve, which are critical to the platform's utility, continue to lack transparency (O'Regan and Choe, 2017; Wachsmuth, 2018; Söderström and Mermet, 2020; Boeing et al, 2021).

The ways in which platforms govern and coordinate their ecosystems of users through opaque uses of data and algorithms is increasingly recognized as producing significant socio-spatial effects, and governance impacts, in cities. Many platform companies such as Airbnb and Uber have deliberately sought to exceed and accelerate their operations beyond the remit of existing governance frameworks, 'moving fast and breaking things' (Stehlin et al, 2020), and in turn provoking regulatory battles worldwide (Cohen, 2018; Wachsmuth and Weisler, 2018; Bainbridge, 2019; Boeing et al, 2021). Such trends diverge from the technology-led urban managerialism characteristic of smart city strategy (Söderström et al, 2014; Vanolo, 2014; McNeill, 2015; Sadowski and Bendor, 2019). As Powell also notes in this collection, the data politics of platforms transcend binary dynamics between 'top-down' authority and 'bottom-up' participation, as everyday digital participation in platform ecosystems continues to amplify multiscalar data accumulation and governance capabilities.

In this sense, the urban implications of digital platforms exceed issues of software design or business models, because they reconstitute contemporary urban socio-spatial relationships in critical ways (Hodson et al, 2021, 26). Everyday spaces of urban encounter and exchange are key sites of intermediation by platform ecosystems, which set up co-constitutive relations of exchange between digitally mediated urban spaces and citizens. Platform urbanism might in one sense be thought of as intimacy at scale: the key here, as Langley and Leyshon put it is, 'for the platform [...] to intermediate the ever-expanding value created by user interactions across their market network' (Langley and Leyshon, 2017, 22). As Törnberg writes (this book), 'platform power thus operates on the interhuman and relational level, seeking to algorithmically modify the social rules that govern social behaviour. Platform power thus implies a relational approach to control, reshaping the connections and relations between people, leveraging social behaviour to generate social pressure for change.'

Platforms implicate questions of scale (see Cinnamon, this collection), as globally expansive platform ecosystems now coordinate huge volumes of urban data and interactions, training and accelerating the use of algorithmically informed urban services across myriad different locales simultaneously. Power is exercised less through centralized 'command and control' operating systems or control rooms, but through more cybernetically realized 'open loops' and feedback mechanisms established through intimately instrumented urban lives and code/spaces (Kitchin and Dodge, 2011; Krivý, 2018; Barns, 2021a). They achieve power through what Mark Graham has called a strategic deployment of 'conjunctural geographies – a way of being simultaneously embedded and disembedded from the space-times they mediate' (2020, 454). Open yet closed; embedded yet disembedded; in place, and no place. 'They can link themselves to the local to [concentrate] reward, and retreat to their ephemeral digital dualisms when abdicating responsibility' writes Graham.

It is in the context of increasingly urbanized platform ecosystems that I want to discuss the emergence of municipal reform policies as a site for policy experimentation. Where the spectre of municipal smart city programmes and centralized urban dashboards emerged as critical focus for urban managerial power and entrepreneurialism, the rapid rise of platform ecosystems has instead provoked cities to respond with novel approaches to data governance, by defending the rights of citizens from wider platform practices of data accumulation and surveillance. Responding to urban platforms, a number of municipal governments actively defend the rights of local workers and residents, championing the need for housing affordability and safe working conditions, and the rights of place-based workers and communities against global platform companies (Domurath, 2018; Wachsmuth and Weisler, 2018; Bainbridge, 2019). As Kitchin has argued in this collection, these different municipal responses to digital disruption can be considered a reflection of the efforts by technology companies to apply market-orientated approaches to governance – whether in partnership with municipal governments through smart city programmes, or through methods to deliberately undermine their capacity to govern effectively. But for custodians of urban spaces, there are clear opportunities for municipal policy actors to redefine conditions of data use and custodianship, redefining how platform ecosystems operate and monetize datafied urban spaces.

As I discuss, these experiments help to reinforce the role of the urban scale in the articulation and renegotiation of progressive data practices. However, they also raise questions about appropriate uses of data power, and its exertion and adoption by both public and private custodians of data, and the relative capacities of citizens and civic institutions to determine how their data should be valued, and for what purposes. Despite their experimental nature, these practices point to the centrality of municipal data politics to the ongoing negotiations of data power across the platform ecosystems of contemporary cities.

Experimental data politics through municipal reform

A number of cities recognize the centrality of urban spaces to the expansion of global platform ecosystems, and have experimented with alternative modes of data engagement that seek to renegotiate terms of platform access to valuable, datafied urban spaces. Most notable among these is the City of Barcelona's digital reform programme, established as part of a wider programme of democratization ushered in with the election of Mayor Ada Colau in 2015. Waterfront Toronto, a government development vehicle that partnered with Google sister company Sidewalk Labs on the now cancelled Sidewalk Toronto precinct, was also active in defining novel forms of data governance to facilitate development of a 'smart', data-driven urban transformation programme in ways that elevated the principle of data stewardship for the public interest (Artyushina, 2020). The Cities Coalition for Digital Rights (CCDR), initially established as a joint initiative between Amsterdam, Barcelona, and New York City in 2018 to promote citizens' digital rights on a global scale (Calzada, 2021), also supports new models of data governance at the municipal level, through both a Digital Rights Governance Project and more recently a Global Observatory of Urban AI. The programme is supported by UN-Habitat under its Digital Cities Toolkit, which aims to put citizens' digital rights at the centre of decision-making about cities' digital transformation programmes, including smart cities programmes.

Each of these programmes seeks to negotiate alternative policies for data use within global platform ecosystems through experimental programmes built around the advocacy of citizen and civic data rights, and in particular the rights of citizens to negotiate how value from their data is being exploited for different applications in the delivery of a range of platform-driven services (Artyushina, 2020; Calzada, 2021). As these examples demonstrate, the capacity for citizen digital rights to be championed and defended by municipal governments rests on broad public acceptance of the role of municipal government as a custodian and champion of citizen data, and is equipped and resourced to ensure privacy protections are effectively maintained. Describing an orientation towards 'platform municipalism', Thompson (2021) has explored the active work across civil society groups to establish new citizen platforms that elevate forms of shared decision-making in place. Linked to the concept of 'new municipalism' (Vollmer, 2017), the local and the municipal is recognized by Thompson as a strategic entry point through which to advance new value frameworks around the community and around concepts of the commons (Gibson-Graham, 2008). As Krisch writes (2022, 60), mobilized through a politics of proximity, platform municipalism mobilizes the urban scale to achieve strategic goals, testing new approaches to collaboration between local authorities, civil society groups, and technology companies and platforms

For the City of Barcelona, the digital reform programme introduced in 2015 ushered in a 'new data deal' designed to revalue urban data not as a commodity but as a fundamental social infrastructure (Morozov and Bria, 2018). Investment programmes in digital capacity building, including new 'data lakes' for data storage and sharing, and new procurement contracts with technology providers, were focused on improving the data literacy and capabilities of the government itself to act as a custodian of citizen data. Here, City's role in relation to technology and data was re-established as a platform not of surveillance and data commodification, but for the integration of a range of public and private data assets to support the achievement of public value outcomes, including the development of new social policies and environmental programmes. Contrasting this approach, in Toronto the potential integration of property development alongside data acquisition by Sidewalk Labs, owned by Google's parent company Alphabet, required a novel approach to data governance resting on the need for data privacy and transparency, articulated in the form of a 'data trust'. In both contexts, the urban setting has played a key role, operating as a locus for commons-based forms of deliberation, protest, and dissent around data rights. As I discuss below, these specific instances of policy experimentation have attracted widespread global interest, demonstrating how municipal experiments help to articulate alternative, appropriate uses of localized citizen data in an era of platform scale. They likewise demonstrate the potentials and indeed 'unexpected persistence' (Mello Rose, 2021) of wider networks of citizen-focused platform infrastructure, which allow multiple cities to coordinate activities and standards in multiscalar ways.

Decidim: an open source platform for municipal democratic reform

Urban data reform in the City of Barcelona represented a 'new data deal' underpinned by novel conditions of data sharing and digital participation in the city (Calzada, 2018; Lynch, 2020; Charnock et al, 2021; Smith and Martín, 2021). Its data reforms manifested a range of civil society pressures experienced within the city as a result of commercial platform ecosystems and their urban impacts. From 2011, under the PSC (the Catalan chapter of the Spanish Socialist Party) Barcelona was a world leader in the advocacy of smart city business opportunities, establishing partnerships with technology providers such as Cisco, Schneider Electric, Accenture, and others, promoting its role as a 'laboratory for the urban future' (March and Ribera-Fumaz, 2019, 232). Occurring simultaneously, during the spring of 2011 the *indignados* or 15-M movement emerged across Spanish cities after the 2008 economic crisis as a series of protests against policies of economic liberalization and austerity (Janoschka and Mota, 2021). Over the following years, Airbnb also became a major short-term accommodation (STA) platform, enabling tourists from across the globe to gain access to residential properties across Barcelona, exacerbating housing shortages and affordability across the city. Through these intersecting movements, the space of the urban became critical to the rearticulation of civic values devoted to supporting the more vulnerable members of society.

Elected in 2015, Mayor Ada Colau emerged as a community leader through her activism against evictions and promotion of affordable housing in Barcelona, under pressure from Airbnb's expansion across the city. Many of her policies were focused on stopping evictions and increasing social housing (Charnock et al, 2021). Colau emerged as a leader of the independent political movement called 'Barcelona en Comú', which was itself formed around a citizen platform called D-CENT, funded as an EU Horizon 2020 project and launched in 2013 by UK think tank Nesta under the leadership of Francesca Bria. The project brought together a number of citizen-led organizations across Europe focused on connecting these organizations with the next generation of open source, distributed, and privacy-aware tools for direct democracy and economic empowerment. D-CENT's establishment coincided with local elections in Spain, seeing an effective alliance between citizen technologists and wider protest movements which saw the election of Barcelona en Comú and, in Madrid, of Ahora Madrid (Calzada, 2018; March and Ribera-Fumaz, 2019).

Barcelona's existing commercial partnerships with smart city providers, most notably Cisco, meant that many of the city's data assets had been privatized under the creation of a 'City OS' data platform. The centrality of D-CENT as a civic platform supporting the popular election of Colau, combined with the significant privatization of the city's digital infrastructure and expansion of Airbnb into the residential housing market, saw the issue of data rights emerge as a focus for reform by the city, which prioritized investment in new digital transformation programmes, pilots, and data-sharing protocols designed to improve the capacity of the citizens to reform government and to in turn advocate on behalf of vulnerable populations. Data and technology issues were elevated as a priority within municipal politics, with key advocates and developers of D-CENT working as civic hackers and technologists now also elevated to key decision-making roles within government. In so doing, urban interventions to create alternate models for data use, governance, and access in cities help to advance what Morozov and Bria called 'a new vocabulary and conceptual apparatus to reassess [cities'] relationship to technology, data, and infrastructures' (Morozov and Bria, 2018, 22). As stated within the City of Barcelona's 2018 data policy:

> The public and private perception of data has to change from that of an asset that offers a competitive advantage to one of a social 'infrastructure' that must be public in order to ensure common well-being, and which is exchanged on a quid pro quo basis.
>
> (City of Barcelona, 2018, 7)

Barcelona's digital reforms attracted widespread interest in the novel form of technopolitics advanced across a city previously championed as a leading smart city, drawing attention to issues of data privacy, technological sovereignty, and data sovereignty as critical issues shaping the city's future (Kitchin et al, 2019; March and Ribera-Fumaz, 2019; Charnock et al, 2021). Progressed under the banner of 'technological sovereignty', the digital reforms elevated the need for self-determination at a municipal level in how a city's data assets should be managed on behalf of its population (Charnock et al, 2021). The D-CENT platform, which had facilitated the rise of Barcelona en Comú as an independent political movement, was in 2016 renamed Decidim (translating as 'we decide') and used by the city to support a participatory engagement programme that allowed citizens to directly vote on budget allocations under the Municipal Action Plan. This allowed voters to 'up-vote' proposals they liked, with some 28,000 voters registered on the platform. One of the most notable of those citizen-led proposals adopted was the iconic (re)design of urban superblocks (*superilles*) to allow for greater liveability and permeability within the city (Aragón et al, 2017; Calzada, 2018).

The city also pioneered new, open source data-sharing contracts with technology companies it contracted to promote access to data assets, rather than allowing these to be privatized through the policies advanced by platform providers. It also introduced a well-publicized pilot for citizen data-sharing called DECODE, which aimed to advance use of cryptographic, privacy-by-design tools to encourage citizens to share their personal data for a variety of public planning purposes (Thompson, 2021, 321). Through DECODE, it was intended that city residents and companies be given the opportunity to 'decide what they keep private and what they want to share, and with whom and under what conditions' (City of Barcelona, 2018, 8). Underpinning these new technology and procurement models were a set of standards and principles that embraced the need for technological sovereignty and digital rights for citizens (Kitchin et al, 2019; Nesta, 2020).

The reform programme also established a set of regulatory conditions that determined how platforms collect and make their data available within the city. A number of private entities were required to essentially 'give back' the data they harvested from Barcelona citizens when accessing their services. Vodafone, the provider of the city's telecoms services, was in this context contractually obliged to share the data it collected from the public, published anonymously on the council's open data website. Here the city government essentially traded the right they gave private operators to access urban space (via licenses and contracts) with the responsibility to share data appropriately on behalf of its citizens, enacting a quid pro quo approach to the governance of digital platforms in the city (Monge et al, 2022).

Ultimately, as a number of reformers active in the development of the data reform programmes have reflected, the capacity for Barcelona's municipal government to advance new data-rights principles reflected the important alliances that existed between civil society groups, civic technologists, and community activists – that is, alliances that existed within civil society beyond ongoing control and administration by government. The founders of Decidim recognized the need to expand beyond Barcelona in order to maintain its independence as an activist platform, establishing a 'MetaDecidim' framework that saw Barcelona become one of a number of different 'instances' of the open source technology (Barandiaran et al, 2018; Calzada, 2018; Smith and Martín, 2021).

The 'Barcelona experiments' attracted attention as a municipal-led programme of data reform. They have been criticized as being overly reliant on using technological fixes to advance reforms around social justice (Charnock et al, 2021). They have, however, also contributed to wider programmes of technology innovation in participatory decision-making at the municipal level. Since the Barcelona reforms were introduced, over 50 additional 'instances' of Decidim have been utilized to support municipal reform programmes worldwide, including the City of Helsinki's citizen-led digital participation programme OmaStadi, and programmes supported in Mérida, Yucatán in Mexico, New York City's City Engagement Commission, and Kakogawa, Japan (Calzada, 2018). Many of these instances reflect work by municipal governments to embed deeper levels of participatory decision-making around how government spending is allocated, facilitating creation, dissemination, and voting of funding proposals directly by citizens (Barandiaran et al, 2018; Smith and Martín, 2021). Other experiments like DECODE remain largely aspirational, as the project struggled to retain the skilled developers needed to bring the project from prototype to platform (Monge et al, 2022). As discussed below, Barcelona also continues to expand its capabilities in the domain of AI governance, advancing experimental new models for AI governance that support improved transparency in the public use of algorithms for city services.

Data trusts for novel urban data governance

Where civic technologists instrumental to the creation of the Decidim platform have advocated a role for municipal government as a custodian of citizen data and champion of data rights, the work undertaken in Toronto in support of a 'data trust' demonstrates a very different set of data politics between citizens, corporations, and municipal government. The Sidewalk Toronto project, generated through an alliance between the public developer Waterfront Toronto and Alphabet subsidiary Sidewalk Labs, proposed that Sidewalk Labs play a key role in building a key waterfront precinct in the city (Quayside) 'from the internet up' (Barns, 2020, 25; Mann et al, 2020). An active and engaged citizenry found limited information being shared about how personal data accessed by Sidewalk Labs as a technology company would be used in its role as urban developer, and faced the spectre of a vertically integrated technology platform that not only controlled data infrastructures but wider urban infrastructures as well. The concept of a 'civic data trust' was consequentially developed in response to a widespread backlash, outlining a novel institutional form to govern how data generated within the Quayside precinct, by diverse platforms including Google, would be used and monetized. The concept of the data trust was designed to recognize the significant urban and public value embodied within an instrumented urban precinct, providing a framework for it to be monetized, and profits shared, across key development stakeholders (Artyushina, 2020; Austin and Lie, 2021).

Central to the concept of the data trust were new definitions around alternative data categories and their data value, including differentiations between 'urban data', such as data from public and shared open spaces, 'conventional data' (or personally identifiable data), and 'transactional data'. The proposal was that the independent data trust would allow urban, conventional, and transactional data to be monetized differently by diverse private companies, but instead of profits being aggregated by one technology company (for example, Sidewalk Labs), these profits could be distributed and shared in diverse ways, including reinvestments in the city's infrastructure. In this experiment, citizens' personal data would continue to be commoditized, but citizens would be able to play a more active role in how they accessed the value of this data; for example, being able to access discounted services and other benefits (Artyushina, 2020, 9), and its monetization would be governed through an independent entity (the trust). The sharing of data value was therefore premised on its ongoing commodification, with citizens invited to participate more actively in its commodification, as compared to the Barcelona experiment in which citizens were invited to share data for public benefits, including future planning and reform of city infrastructure.

The Sidewalk Toronto programme was cancelled by Waterfront Toronto in March 2020, following a successful activist campaign under the #BlockSidewalk movement that targeted issues around data governance premised within the data trust model. It has attracted widespread attention as a failed smart city project which highlights the importance of maintaining a 'social licence to operate' where there is public confidence in the development of experimental technologies and services at the urban scale (Artyushina, 2020; Mann et al, 2020; Austin and Lie, 2021). The failure of the civic data trust model advanced under the project reflects the highly contested conditions through which different forms of value-sharing around data can be negotiated, particularly in relation to privacy and surveillance concerns (Austin and Lie, 2021). Key questions remain about the appropriate institutional form – whether a novel data trust, a municipal government, or a civil society platform – capable of determining standards and protocols for different kinds of value embedded in different kinds of data, whether public or 'open', transactional, or personally identifiable. As Artyushina has argued, if data is perceived primarily as a financial asset, then a data trust model established as a quasi-collective institutional form to manage how this value is shared by producer and consumers of data remains constrained, because it ultimately seeks to advance the privatization of urban data governance and entrench practices of data surveillance (Artyushina, 2020, 10).

Data rights principles and protocols: the Cities Coalition for Digital Rights

For these reasons, the work of the Cities Coalition for Digital Rights (CCDR) represents an important development in the articulation of novel approaches to data governance at the urban scale. Unlike the proposed data trust model outlined through the ill-fated Toronto example, the CCDR, in partnership with UN-Habitat and with support from localized municipal governments, enacts and advocates for principles and standards for the adoption and defence of digital rights at the municipal scale spanning a number of different categories for digital rights.

Initially formed in 2018, the CCDR has remained active in supporting the work of Barcelona and other European cities in developing new standards for urban data sharing, and now has over 45 cities as signatories to a set of digital rights principles. The five digital rights principles championed by the CCDR include: (i) right to universal and equal access to the internet, and digital literacy; (ii) right to privacy, data protection, and security; (iii) right to transparency, accountability, and non-discrimination of data, content, and algorithms; (iv) right to participatory democracy, diversity, and inclusion; and (v) right to open and ethical digital service standards (CCDR, nd). As Calzada reflects in relation to these digital rights principles, it is clear that an agenda to promote digital rights must recognize that these are not a set of rights in and of themselves, but are also related to other human rights, particularly freedom of expression and the right to privacy in online and digital environments (2021, 5).

In 2022, in partnerships with the City of Barcelona, London, Amsterdam, and UN-Habitat, CCDR launched a Global Observatory of Urban Artificial Intelligence, which extends its advocacy of digital rights to AI initiatives at the local scale. The CCDR also identified cities as increasingly experimental sites for new forms of artificial intelligence and automation technologies being applied across a range of sectors, from predictive policing to environmental monitoring and beyond. It advocates for improved governance frameworks and principles that allow municipal governments to act as effective custodians of public spaces, as they become an increasing focus of technology companies seeking to test out and expand AI operations. It acts as a wider network champion, allowing experiments and case studies in individual cities to be translated to other urban contexts. One example of this work includes the 2020 establishment of an AI register in Helsinki and Amsterdam, designed to enable citizens to gain a better understanding of how algorithms are being used within municipal government, and for what purposes, in a transparent way.

Another initiative promoted through the CCDR is an 'AI life cycle' tool, designed to support city governments to evaluate different forms of risk at different stages of AI's implementation, from framing and design, including data collection and problem identification, to implementation, deployment, and maintenance. As Calzada (2021) argues in a survey of the activities, the work of the CCDR points to a role for transnational agencies and networks in advocating for novel policy experiments and standards, working in partnership with municipal reform partners to test challenges and opportunities at the urban scale in a collaborative rather than competitive way. As Calzada also raises, the capacity for cities engaged with the CCDR to progress digital rights frameworks appears to depend fundamentally on the presence of an active civil society, in which citizen groups, associations, and NGOs work collaboratively to push stakeholders towards more inclusive data governance models at the municipal level (2021, 19).

As these examples demonstrate, policy experiments that seek to articulate alternative forms of data governance are proliferating at the municipal scale, in ways that respond to the data practices of platform ecosystems. What is clear, however, is that this work depends on relatively nascent conceptions of data rights and alternative forms of data value, particularly in relation to the role of municipal governments in defending wider data values and rights on behalf of citizens. In his reflection on the capacities of municipal governments to enact new digital rights frameworks, Calzada (2021, 4) notes that contemporary citizenship is being implicated by digital settings across a host of different dimensions, including 'access, openness, net-neutrality, digital privacy, data encryption, protection and control, digital/data/technological sovereignty'. In turn, as these examples show, there are many different emergent digital rights 'taxonomies' that ultimately incorporate a mix of concepts, principles, and expectations around how the value of data, spanning personal, public, or 'open data', environmental data, transactional data, health data, and so on, should be protected but also exploited by different actors working towards public, civil society, and commercial outcomes.

As such, questions remain not only about what digital rights are to be championed at the municipal scale, but also the appropriate instruments needed to ensure their protection. What principles should govern the use and exploitation of personal data by reforming municipal government seeking to deliver public services on behalf of citizens, including, for example, local health programmes? Who are the most critical stakeholders in relation to the articulation and negotiation of novel data rights frameworks? Should these negotiations be undertaken in partnership with existing platform ecosystems? How can city alliances support broader policy reforms in city data use and sharing by different institutional partners? Should governments advance the right of citizens to exploit commercially the value of their data if platform companies do this? Such questions have driven a number of policy experiments to date, with inconclusive results. Clearly there will be future campaigns and novel experiments that will continue to grapple with such questions in the future, particularly in relation to the proliferation and governance of AI platforms in cities. Challenges around appropriate institutional forms for data custodianship remain central to many of these experiments and their capacity to scale over time, suggesting the need for ongoing work not only to test and experiment with novel citizen platform models, whether municipal data trusts, data lakes, and encrypted data platforms, but also closer attention to the principles and frameworks underpinning governing how data is valued, for whom, and by whom.

Conclusion

If the 2010s saw the discursive positioning of platform ecosystems as disruptors of urban services and enablers of 'open innovation' across cities, the 2020s is emerging as a critical decade for the renegotiation of urban data governance, in ways that signal the enduring role of (datafied) urban spaces as sites for the contestation of diverse 'rights to the city'. The global rise of platform ecosystems has, to date, facilitated extractive conditions of data access and use in ways that advance forms of algorithmic data enclosure (Andrejevic, 2009), while being positioned as enabling innovation in the delivery of more responsive, smart, and adaptive services (Söderström et al, 2014; Shelton et al, 2015; Sadowski and Bendor, 2019). These conditions of data capture have in turn provoked new municipal reforms that seek to articulate alternative models of data governance, by prioritizing citizen data rights, and improving literacies and skills in the custodianship and management of urban data on behalf of urban citizens.

These new directions are a reminder of the agency of municipal governments in wider global debates around surveillance capitalism and the platform society, and serve to underscore the ongoing criticality of the urban scale as a site of struggle for the renegotiation of conditions of equity, justice, and fairness in the governance of the increasingly datafied code/spaces of the city. As Graham (2020, 456) has argued:

> Geography clearly does matter, but not simply as a way to describe the tethering of platforms to places. It is rather in the conjuncture of tethered and untethered relationships with space that we need to envision how platforms bring new digital geographies into being– and envision how we can tame them.

If municipal experiments around new data politics are highly local, contingent, temporary, and provisional – indeed, even if they fail to scale – perhaps this is a reminder of the vitality of the urban as a site of struggle and agency within the urban-digital spaces of contemporary platform ecosystems we find ourselves living with.

Bibliography

Andrejevic, M. (2009) 'Privacy, exploitation, and the digital enclosure', *Amsterdam Law Forum*, 1(4): 47–62.

Aragón, P., Kaltenbrunner, A., Calleja-López, A., Pereira, A., Monterde, A., Barandiaran, X.E. and Gómez, V. (2017) 'Deliberative platform design: The case study of the online discussions in Decidim Barcelona', *Lecture Notes in Computer Science*, 10540: 277–87.

Armstrong, M. (2006) 'Competition in two-sided markets', *The RAND Journal of Economics*, 37(3): 668–91.

Artyushina, A. (2020) 'Is civic data governance the key to democratic smart cities? The role of the urban data trust in Sidewalk Toronto', *Telematics and Informatics*, 55. https://doi.org/10.1016/j.tele.2020.101456

Austin, L. and Lie, D. (2021) 'Data trusts and the governance of smart environments: Lessons from the failure of Sidewalk Labs' Urban Data Trust, *Surveillance and Society*, 19(2): 255–61.

Bainbridge, A. (2019) 'Uber "came to our shores, illegally, like pirates", class action lead plaintiff says', ABC News, [online] 2 May, available from: https://www.abc.net.au/news/2019-05-03/uber-to-face-class-action-against-taxi-and-private-drivers/11073640

Barandiaran, X., Calleja-López, A. and Monterde, A. (2018) 'Decidim: Political and technopolitical networks for participatory democracy', Decidim's project white paper, [online], available from: http://ajbcn-meta-decidim.s3.amazonaws.com/uploads/decidim/attachment/file/2005/White_Paper.pdf

Barns, S. (2020) *Platform Urbanism: Negotiating Platform Ecosystems in Connected Cities*. London: Palgrave Macmillan.

Barns, S. (2021a) 'Out of the loop? On the radical and the routine in urban big data', *Urban Studies*, 58(15): 3203–10. https://doi.org/10.1177/00420980211014026

Barns, S. (2021b) 'Joining the dots: Platform intermediation and the recombinatory governance of Uber's ecosystem', in M. Hodson, J. Kasmire, A. McMeekin, J.G. Stehlin and K. Ward (eds) *Urban Platforms and the Future City: Transformations in Infrastructure, Governance, Knowledge, and Everyday Life*, Abingdon: Routledge, pp 87–104.

Boeing, G., Besbris, M., Wachsmuth, D. and Wegmann, J. (2021) 'Tilted platforms: Rental housing technology and the rise of urban big data oligopolies', *Urban Transformations*, 3. https://doi.org/10.1186/s42854-021-00024-2

Calzada, I. (2018) '(Smart) citizens from data providers to decision-makers? The case study of Barcelona', *Sustainability*, 10(9). https://doi.org/10.3390/su10093252

Calzada, I. (2021) 'The right to have digital rights in smart cities', *Sustainability*, 13(20). https://doi.org/10.3390/su132011438

CCDR. (nd) 'Our principles and declaration', Cities Coalition for Digital Rights, [online], available from: https://citiesfordigitalrights.org/thecoalition

Charnock, G., March, H. and Ribera-Fumaz, R. (2021) 'From smart to rebel city? Worlding, provincialising and the Barcelona Model', *Urban Studies*, 58(3): 581–600. https://doi.org/10.1177/0042098019872119

Choudary, S.P. (2015) *Platform Scale: How an Emerging Business Model Helps Startups Build Large Empires with Minimum Investment*, Singapore: Platform Thinking Labs.

City of Barcelona. (2018) Government Measure on Ethical Data Management, [online] May, available from: https://www.barcelona.cat/digitalstandards/en/data-management/0.1/summary

Cohen, T. (2018) 'Being ready for the next Uber: Can local government reinvent itself?', *European Transport Research Review*, 10. https://doi.org/10.1186/s12544-018-0330-8

Domurath, I. (2018) 'Platforms as contract partners: Uber and beyond', *Maastricht Journal of European and Comparative Law*, 25(5): 565–81. https://doi.org/10.1177/1023263X18806485

Fowler, G.A. (2015) 'There's an Uber for everything now', *Wall Street Journal*, [online] 5 May, available from: https://www.wsj.com/articles/theres-an-uber-for-everything-now-1430845789

Gibson-Graham, J.K. (2008) 'Diverse economies: Performative practices for "other worlds"', *Progress in Human Geography*, 32(5): 613–32.

Gillespie, T. (2010) 'The politics of platforms', *New Media and Society*, 12(3): 347–64.

Graham, M. (2020) 'Regulate, replicate, and resist – the conjunctural geographies of platform urbanism', *Urban Geography*, 41(3): 453–57. https://doi.org/10.1080/02723638.2020.1717028

Helmond, A. (2015) 'The platformization of the web: Making web data platform ready, *Social Media + Society*, 1(2). https://doi.org/10.1177/2056305115603080

Hodson, M., Kasmire, J., McMeekin, A., Stehlin, J.G. and Ward, K. (eds) (2021) *Urban Platforms and the Future City: Transformations in Infrastructure, Governance, Knowledge and Everyday Life*, Abingdon: Routledge.

Janoschka, M. and Mota, F. (2021) 'New municipalism in action or urban neoliberalisation reloaded? An analysis of governance change, stability and path dependence in Madrid (2015–2019)', *Urban Studies*, 58(13): 2814–30. https://doi.org/10.1177/0042098020925345

Kitchin, R. and Dodge, M. (2011) *Code/Space: Software and Everyday Life*, Cambridge, MA: MIT Press.

Kitchin, R., Cardullo, P. and Di Feliciantonio, C. (2019) 'Citizenship, justice and the right to the smart city', in P. Cardullo, C. Di Feliciantonio and R. Kitchin (eds) *The Right to the Smart City*, Bingley: Emerald, pp 1–24.

Krisch, A. (2022) 'From smart to platform urbanism to platform municipalism: Planning ideas for platforms in Toronto and Vienna', in A. Struver and S. Bauriedl (eds) *Platformization of Urban Life: Towards a Technocapitalist Transformation of European Cities*, Berlin: De Gruyter, pp 53–71.

Krivý, M. (2018) 'Towards a critique of cybernetic urbanism: The smart city and the society of control', *Planning Theory*, 17(1): 8–30.

Langley, P. and Leyshon, A. (2017) 'Platform capitalism: The intermediation and capitalization of digital economic circulation', *Finance and Society*, 3(1): 11–31.

Lynch, C.R. (2020) 'Contesting digital futures: Urban politics, alternative economies, and the movement for technological sovereignty in Barcelona', *Antipode*, 52: 660–80. https://doi.org/10.1111/anti.12522

Mackenzie, A. (2018) 'From API to AI: Platforms and their opacities', *Information, Communication & Society*, 22(13): 1989–2006. https://doi.org/10.1080/1369118X.2018.1476569

Mann, M., Mitchell, P., Foth, M. and Anastasiu, I. (2020) '#BlockSidewalk to Barcelona: Technological sovereignty and the social license to operate smart cities', *Journal of the Association for Information, Science and Technology*, 71(9): 1103–15.

March, H. and Ribera-Fumaz, R. (2019) 'Barcelona: From corporate smart city to technological sovereignty', in A. Karvonen, F. Cugurullo and F. Caprotti (eds) *Inside Smart Cities: Place, Politics and Urban Innovation*, London: Routledge, pp 229–42.

McNeill, D. (2015) 'Global firms and smart technologies: IBM and the reduction of cities', *Transactions of the Institute of British Geographers*, 40(4): 562–74. https://doi.org/10.1111/tran.12098

Mello Rose, F. (2021) 'The unexpected persistence of non-corporate platforms: The role of local and network embeddedness', *Digital Geography and Society*, 2. https://doi.org/10.1016/j.diggeo.2021.100020

Monge, F., Barns, S., Kattel, R. and Bria, F. (2022) 'A new data deal: The case of Barcelona', UCL Institute for Innovation and Public Purpose, Working Paper Series (WP 2022/02), available from: https://www.ucl.ac.uk/bartlett/public-purpose/wp2022-02

Morozov, E. and Bria, F. (2018) *Rethinking the Smart City: Democratizing Urban Technology*, New York: Rosa Luxemburg Stiftung.

Nesta. (2020) 'Common knowledge: Citizen-led data governance for better cities', London: Nesta.

O'Regan, M. and Choe, J. (2017) 'Airbnb and cultural capitalism: Enclosure and control within the sharing economy', *Anatolia*, 28(2): 163–72. https://doi.org/10.1080/13032917.2017.1283634

Pollio, A. (2021) 'Uber, airports, and labour at the infrastructural interfaces of platform urbanism', *Geoforum*, 118: 47–55. https://doi.org/10.1016/j.geoforum.2020.11.010

Raetzsch, C., Pereira, G., Vestergaard, L.S. and Brynskov, M. (2019) 'Weaving seams with data: Conceptualizing City APIs as elements of infrastructures', *Big Data & Society*, 6(1). https://doi.org/10.1177/2053951719827619

Sadowski, J. (2021) 'Who owns the future city? Phases of technological urbanism and shifts in sovereignty', *Urban Studies*, 58(8): 1732–44. https://doi.org/10.1177/0042098020913427

Sadowski, J. and Bendor, R. (2019) 'Selling smartness: Corporate narratives and the smart city as a sociotechnical imaginary', *Science, Technology, and Human Values*, 44(3): 540–63. https://doi.org/10.1177/0162243918806061

Shelton, T., Zook, M. and Wiig, A. (2015) 'The "actually existing smart city"', *Cambridge Journal of Regions, Economy and Society*, 8: 13–25.

Simon, P. (2011) *The Age of the Platform: How Amazon, Apple, Facebook, and Google Have Redefined Business*, Henderson, NV: Motion Publishing.

Smith, A. and Martín, P.P. (2021) 'Going beyond the smart city? Implementing technopolitical platforms for urban democracy in Madrid and Barcelona', *Journal of Urban Technology*, 28(1–2): 311–30. https://doi.org/10.1080/10630732.2020.1786337

Söderström, O. and Mermet, A.-C. (2020) 'When Airbnb sits in the control room: Platform urbanism as actually existing smart urbanism in Reykjavík', *Frontiers in Sustainable Cities*, 2. https://doi.org/10.3389/frsc.2020.00015

Söderström, O., Paasche, T. and Klauser, F. (2014) Smart cities as corporate storytelling. *City: Analysis of Urban Change*, 18(3): 307–20. https://doi.org/10.1080/13604813.2014.906716

Stehlin, J., Hodson, M. and McMeekin, A. (2020) 'Platform mobilities and the production of urban space: Toward a typology of platformization trajectories', *Environment and Planning A: Economy and Space*, 52(7): 1250–68. https://doi.org/10.1177/0308518X19896801

Strüver, A. and Bauriedl, S. (eds) (2022) *Platformization of Urban Life: Towards a Technocapitalist Transformation of European Cities*, Berlin: De Gruyter.

Swyngedouw, E. (2005) 'Governance innovation and the citizen: The Janus face of governance beyond-the-state', *Urban Studies*, 42(11): 1991–2006.

Thompson, M. (2021) 'What's so new about New Municipalism?', *Progress in Human Geography*, 45(2): 317–42. https://doi.org/10.1177/030913252 0909480

Tiwana, A. (2013) *Platform Ecosystems: Aligning Architecture, Governance, and Strategy*, Amsterdam: Morgan Kaufmann.

Vanolo, A. (2014) 'Smartmentality: The smart city as disciplinary strategy', *Urban Studies*, 51(5): 883–98. https://doi.org/doi:10.1177/004209801 3494427

Vollmer, L. (2017) 'Keine Angst vor Alternativen. Ein neuer Munizipalismus. Über den Kongress "FearlessCities", Barcelona 10./11'', *Suburban: Zeitschrift für Kitische Stadtforschung*, 5(3): 147–56.

Wachsmuth, D. and Weisler, A. (2018) 'Airbnb and the rent gap: Gentrification through the sharing economy', *Environment and Planning A: Economy and Space*, 50(6): 1147–70. https://doi.org/10.1177/0308518X18778038

7

Webinars and War Rooms: Technopolitics of Data in the Digitalizing State

Ayona Datta and Ola Söderström

Introduction

During the early days of the COVID-19 pandemic in India, the Indian Smart Cities Mission (SCM) underwent a radical policy shift. In March 2020, right after a nationwide lockdown was put in place, the mission director and joint secretary of SCM (previously CEO of another Indian smart city), alongside various smart city CEOs, organized a series of webinars around the role of smart cities in tackling the pandemic. A key aspect of these webinars was to exchange knowledge across the various smart cities in terms of how the pandemic could be controlled and managed using the smart technologies already commissioned in these cities. The webinars also sought to disseminate information to the public and engage in discussion about the future of these smart cities. These Smart Cities webinars themselves were not a new initiative. Indeed, webinars were a regular feature on social media where stakeholders and consultants to the SCM would be invited to present different aspects of smart technologies from time to time. Yet it was during the pandemic that these webinars took on a new meaning and significance as the face of the state. As the mission director noted:

> 'Technology isn't only to be seen in the light of digital technology. It cannot be seen in isolation and has to be seen with various different things. We can call it digilogue- which is digital+analogue, where analogue is a lot more fluid. And they need to come together.'
>
> (Tech Clinic Webinar 1)

The mission director as an actor of the federal Ministry of Human and Urban Affairs (MoHUA) presents the SCM as a real Latourian 'centre of calculation' (Latour, 1987) in a country under lockdown. These webinars were led by him and his colleagues to simultaneously frame the policy at national level and centralize the experience of each smart city in a rhetoric and practice of federal patronage. These webinars were also significant in the production of 'war rooms' – existing Integrated Command and Control Centres (ICCC) of Smart Cities repurposed into 'weapons' for managing and tackling the spread of COVID-19. The war rooms presented a military semantic field of a series of strategies and tools akin to what Morozov (2013) has noted as a 'technological solutionism' towards a virus seen to be unmanageable through public health initiatives alone.

Our chapter draws on these webinars as rich material for understanding the 'digitalising state' (Datta, 2023) in a crisis-centric mode of existence (Söderström, 2021). We consider these online events as sites of 'technopolitics in the making' through the narrative and visual exchanges between different levels and institutions of the state and private stakeholders. We refer to technopolitics, following Hecht, as 'strategic practice of [...] using technology to constitute, embody, or enact political goals' (Hecht, 2009, 15). The term technopolitics is in our view useful to think about technologies as neither determining political outcomes nor simply determined by political strategies. It helps to consider technologies as both offering affordances for political action and being shaped by political strategizing. The discourse on the power of data and the spectacle of data through digital visualizations are, this chapter shows, central to the technopolitical strategies developed by the Indian state during the most critical moments of the COVID-19 pandemic.

We analyse the social and mediatic space of these webinars in terms of what they aimed to do, who participated, and the 'smartness mandate' (Halpern et al, 2017) upheld as a collective discourse in these webinars. We first define the specific technopolitics of the Indian state during the pandemic crisis, and show how it was engineered by repurposing the ICCCs into war rooms and performed through a series of webinars. We then focus on four main articulations of technopolitics in these webinars: performative imaginaries of data in times of crisis; war rooms as spaces of technopolitical scale orchestration; uneven and selective geographies of data; and data resilience without privacy. We conclude by reflecting on how Indian data politics, both highly centralized in its organization and much more fragmented across scales and actors when observed in action, blurs the idea of a frictionless roll-out of the smartness mandate.

Technopolitics and data power in the (un)making of a crisis

Studies of technopolitics show that they can have a centralized character when there is a strong statal programme or project; for instance, in the case of the

French nuclear energy programme. However, technopolitics is never simply a form of statecraft. Other forms of technopolitics, such as 'citizen sensing' (Gabrys, 2014) or 'citizen science' (Haklay, 2013) illustrate that technopolitics is always a distributed construction, an assemblage of technological and political imaginaries. Technopolitics are crafted by discourses and material elements: rationalities and technologies in the Foucauldian idiom. They use the support of what Jasanoff and Kim (2013) have called 'socio-technical imaginaries', expressed in discourse or rationalities. They are also tethered to concrete technologies. These entanglements of words and things is a significant characteristic of technopolitics in the making as we observed in the webinars.

As Halpern and Mitchell (2023) argue, while technopolitics can be framed by various discourses and metanarratives, smartness is a specific mandate: a promise related to the deployment of computation. It is an epistemology constructed by the articulation of different lineages of technologies and rationalities finding their roots in Cold War rationalities of the 20th century and inextricably linked to the language of crisis (Halpern and Mitchell, 2023). For them, there is no big plan behind the smartness mandate. It is a way of thinking that connects spaces, populations, and algorithmic management to ensure resilience. It is a way of seeing that searches for opportunities in crises.

We argue that in India the smartness mandate comes not from the Cold War genealogies (albeit the technological dependency on global IT corporates has connections to global geopolitics as well) but mainly from a genealogy of Indian modernity in which technology has perpetuated a myth of progress and modernity since the making of the Indian nation. This works much like what Donna Haraway (1988) suggests as the 'god trick' of technology. She notes (p 581), 'Vision in this technological feast becomes unregulated gluttony; all seems not just mythically about the god trick of seeing everything from nowhere, but to have put the myth into ordinary practice'. In India, technology and mythology are often interchangeable since technology perpetuates a mythology of modernity, and myth-making around nationhood and modernity is used to magnify the need for technology as an end in itself. Datta (2019) has noted that the overlapping narratives of a mythical Hindu nation in India are bolstered by the techno-utopian imaginaries of modernity. So while mythology is an ideology and technology is a supposed 'rationality', technology and mythology are in fact complementary narratives in India's march to progress since independence in 1947.

However, transitioning to a digital society has not been smooth in India. Just as Isin and Ruppert (2015) have noted, the state engages in simultaneous 'callings' and 'closings' to convince the public to move into digital space. Callings are forms of (soft and hard) state coercion through which citizens are compelled to sign on to digital platforms and engage as digital citizens, while closings are the simultaneous redlining of information and the freedom of citizens to contribute in democratic ways over the internet and many other digital platforms. In India too, Sukumar notes that the acceptance of technology among the masses was far from simple. He argues that 'successive governments in New Delhi have realized that the relationship between technology and citizen cannot be easily mediated, especially in an open and diverse society such as India's.' (Sukumar, 2019, xxviii). We see this continuing in the more recent Indian 100 Smart Cities Mission where Datta (2018) noted that the Indian state engaged in strategies of enumeration and articulation of digital citizenship through a top-down approach to convince citizens to get onboard its smart cities policies.

This is significant in the context of war rooms where technology performs a 'god trick' of control while unleashing a coercive force on citizens to use their smartphones and monitor their own health via apps, resulting in a data harvest for the state and the private companies that collect this data. As Datta et al (2021, 382–3) argue, the 'Apps, Maps and War rooms' produced during the COVID crisis determine a field of COVtech – 'which ranges from the most aggressive war room-based ubiquitous surveillance to text-based information dissemination. [...] Incorporating technologies of early 20th-century plague cartography within current Smart City surveillance machines, COVtech lends itself to both map-based and app-based surveillance of the virus present in human bodies.' It is COVtech that we see articulated in the webinars through the mythologies of control and order embodied in the various technologies of the war room. This makes the webinars a space where the myth is validated through the performative dynamics between state, private consultants, and civil society at large, as well as the space where the soft power of the state is established through technology to its citizens. We further argue that this myth is both enabling as well as limiting, for while it mobilizes increased communication between the state, its stakeholders, and civil society in the war rooms, it also produces some 'guarded invisibilities' (Furlong, 2020) around data itself. The webinars we argue maintain strategic invisibilities around data use, storage, and control while performing its visibility as a war machine in the service of citizens.

Repurposing integrated command and control centres into war rooms

If 'control room' signifies a particular combination of architecture and hardware, it also, in a single phrase, signifies a meshing of jurisprudence, communications, media technology, networks, sovereignty and space.
(Deane 2015, np)

Command and control centres have received increased attention among recent science and technology studies (STS) scholars. As Deane (2015) argues, a long genealogy of technological development of the digitalizing state produced the command and control rooms as a 'techno-aesthetic' form of power and control that obscures its contentious links to state surveillance. Mapping 'new configurations of power beyond complexity' in the control room, Deane cautions against 'the grammatizing dream of information and the information sciences, which is the promise of pure remote control and an end to the encumbrance of materiality' (np).

More recently geographers have become interested in the limitation and inadequacies of a cybernetic and engineering approach, insufficient to tackle the 'wicked problems' (Goodspeed, 2015) of urban transformations. In particular, their apparent neutrality and factuality have been questioned by various scholars. Mattern (2015) has noted that the visualization of information in ICCCs has its origins in airplane cockpits, arguing that 'these urban dashboards perpetuate the fetishization of data as a "monetizable" resource and a positivist epistemological unit – and they run the risk of framing the city as a mere aggregate of variables that can be measured and "optimized" to produce an efficient or normative system' (np). For Kitchin et al (2016, 95), who conducted a critical analysis of dashboards based on their design of the Dublin Dashboard, these 'are inherently active and ideological. They express a particular vision of cities and urban governance; a normative notion about what should be measured, what should be asked, and what should be revealed; and they have normative effect, shaping decision-making and behaviour'.

Particularly relevant for our analysis is Luque-Ayala and Marvin's (2020) study of the first ever Smart Operations Centre in Rio de Janeiro. In their reading, ICCCs are operational in the management of flows and circulation, which constitutes in their view one of seven computational logics they identify in contemporary cities. They constitute a meta-infrastructure in the sense that they assemble and coordinate a series of otherwise disconnected network of digitalized infrastructures. Fundamentally, they argue, ICCCs perform a 'specific form of governmentality'. This form of governmentality cannot be understood as a panoptic 'black-boxed' surveillant *dispositif* aiming to discipline bodies. It is rather 'based on the need to continuously adapt to conditions of turbulence via the operationalization of infrastructural flows and the development of novel ways of seeing and engaging with the city and its infrastructures' (p 152). ICCCs have thus become emblems of 'the' smart city, and flagships, as the buildings hosting them visualize in space the ambition of a 'smart' municipality: 'they function as materializations of smart city visions produced by city authorities as well as corporate and other actors' (Caprotti, 2019, 4). Starting from the earliest ICCCs to the more recent ICCCs unfolding in smart cities across the world, the development of visual aesthetics of ICCCs have contributed to the wider discourse and performance of COVtech in the Indian war rooms.

There are, however, several aspects where Indian ICCCs depart from Luque-Ayala and Marvin's conclusions, notably concerning the minor role of surveillance and processes of (un-)blackboxing, where infrastructures and data would be in the public view. In India, Praharaj (2020) observed that although the primary motivation for the ICCCs was to integrate decision-making across all state levels and departments, their lack of digital capacity meant that 'the command-and-control centres in Indian smart cities are predominantly privatized and there is an inclination towards big data corporatization' (p 79). In their research on smart cities in India, Datta et al (2021) further noted that the smart technologies in these ICCCs were being outdated even as new tenders were created for their installation, and as several new ICCCs were mired in delays in recruiting project consultants and technology partners. Further, in smaller cities bureaucracies across state institutions and between federal and municipal scales significantly slowed down transfer of technology from private sector partners to the state. Praharaj (2020) also found that only 'an average 7% of the total smart city investment was allocated for building the big data centres by the 83 out of 100 municipalities that undertook such developments' (p 85). These differences are, we believe, a testimony of the various existing forms of 'provincialization' (Chang et al, 2020) of smart urban technologies and data politics in the Global South; that is, how international smart city policies and technological standardizations land in local contexts of the Global South and produce fragmented and incomplete versions of their aspirations for ubiquitous surveillance and urban managerialism.

In India, ICCCs must also be understood in the context of their re-engineering and repurposing as 'COVID war rooms' during the COVID-19 public health emergency. 'War rooms' in computing are spaces where key people get together to solve crises via improved workflows, bringing together people who would not necessarily collaborate with one another. Although war rooms designate meetings where urgent action is taken, in the context of COVID-19, war rooms took on new significance as militaristic operations to stop the spread of a virus. As Datta et al (2021) note 'The war rooms do not produce new machines per se, rather consolidate and concretize a decade of preparatory work done by urban municipalities in mapping each city to make visible and give meaning to the data harvest from COVID19-related apps and maps' (p 385). Thus, the war rooms visualized the experiments already underway with a range of smart technologies, integrating them within the ICCCs and discursively placing them in the mode of COVtech.

The webinars reveal the increasing value and significance of ICCCs during the pandemic and their role in Indian administration into the future. This is significant, since under the SCM guidelines, although cities could decide their own customized package of smart technologies, the ICCCs were the common mandatory initiative across all cities. Developed and assembled by a host of international and domestic IT partners, infrastructure companies, and private sector consultants, ICCCs produce an operational mechanism that ostensibly connects data to governance, administration, and jurisprudence. They are now established across 83 of the 100 smart cities, although some are still a long way away from the real-time features of a conventional ICCC.

A focus on webinars and war rooms highlights how the smartness mandate in India is also a political mandate for a longer-term legacy of 'technological solutionism' (Morozov, 2013) to endure through future decades. In the war rooms, the discourse of technology functions, as Zook and Spangler (2023) argue, as 'a discursive construction that creates suitable conditions for the manufacture and extraction of data as an asset' (p 110). They serve a dual purpose of state expansion through data colonialism – of extraction and capitalization of citizen data. As Eubanks (2018) notes, these war rooms 'automate inequality' through the quantification of crisis, risk, and resilience.

Webinars as method

In this project we have taken webinars as a research 'field' in which to observe a technopolitics in the making of COVID war rooms. Webinars can appear mundane for data collection, but they can be seen as places where technopolitics unfolds during COVID-19. The unprecedented nature of the webinars where the state enacts a public performance of governance – indeed, recording this for posterity – makes the webinars forms of artefacts that are non-textual, visual, and a 'thicker' articulation of the processes that occurred behind closed doors in the pre-pandemic era. Webinars are part of the 'online turn' that became the hallmark of the COVID-19 pandemic and became the 'face' of the state. This meant that as the restrictions of safe distancing and lockdowns led to a rise in online meetings, the state itself moved online to YouTube/Facebook live meetings between state officials and stakeholders. These online meetings facilitated an unprecedented presence of the state in social media, exchanging knowledge across urban and federal levels and with private-sector partners nationally and globally.

The webinars were the virtual spaces where the guidelines of the redeployment from ICCCs to war rooms was imagined, implemented, showcased, and coordinated across Indian smart cities. The webinars were products of a strong statal programme, accelerated and legitimized by the pandemic crisis in 2020. But they were also the result of a range of distributed actors as well as of processes of policy learning and policy mobility across cities in and outside India, which we observed in action in the webinars. These webinars became the mediated spaces of performative engagement between the state and 'para-state' – contractors brought into close contact with bureaucracies of the state through digital infrastructures – which supports a 'new world of big governments, IT functions and their relationships with big service providers' (Dunleavy et al, 2006). Through these new relationships with the para–state, the state enacted what Meijer (2018) calls 'governance games', which includes a 'politics of data collection, data storage, data usage, data visualization and data access' (Meijer, 2018, 195). Here the performative space of webinars was rendered as rational and neutral with glitzy visualizations and animations of the spread of disease presented by civil servants, national and international professionals providing their services to the Smart Cities Mission.

The webinars were organized by federal government and parastatal institutions such as the National Institute of Urban Affairs (NIUA), Smart Cities Mission (SCM), and the Ministry of Human and Urban Affairs (MoHUA), which were mainly responsible for managing the COVID-19 response of the state. These were attended from time to time by Smart City CEOs from different Indian cities who were invited to share their experiences of managing the COVID-19 crises in their respective cities in order to facilitate knowledge exchange and mutual learning. Private IT consultants (both national and international) also featured heavily as the repositories of knowledge and expertise in boosting the technological capacity of the state in managing COVID-19. Prominent among these were Microsoft, ESRI, KPMG, and Scanpoint, each of which had specific roles in delivering technologies during the pandemic.

Using technology provided by private companies, the presenters demonstrated the ways in which services – ranging from heat maps to contact tracing, from delivery of essential items to citizens' engagement – would address problems that came up locally and otherwise. Occasionally CSOs and think tanks (such as Internet Democracy) would also be invited, who would advocate for data rights and justice for ordinary citizens. The majority of the issues presented by the speakers revolved around how smart cities shall tackle the epidemics, and how smart city technologies have been put to use in dealing with such situations. These webinars looked at the experiences of various municipalities (Chandigarh, Bengaluru, Pune, Varanasi) in leveraging existing digital technology for COVID-19 management along with the use of emerging technologies for post-COVID-19 realignment that is likely to take place across the country. Applications used by different war rooms were mostly similar, but some states have deployed these technologies based on the need of that city – more intensively. These can be categorized into four purposes: information, communication, management, and preparedness.

The webinars were mostly steered by the SCM director at the federal level, who chaired most of these events, leading the direction of the discussion as well as shaping the agenda for subsequent webinars. As SCM director, he was the official spokesman of the federal state; yet at the same time he was also learning from the presentations of other regional and urban state officials. Through the course of the webinars, he directed the sharing of experiences and knowledge across cities and levels of government, enabled stakeholders to showcase various local initiatives, and outlined potential directions of using COVID related technologies in the future of urban governance.

In using webinars as method, we focus on the discourses that underpin these exchanges within the webinars. We examine in the Foucauldian sense of the word – recurring articulated arguments – an imagined future that is pandemic-free. This discourse is shaped by 'socio-technical imaginaries', to use Jasanoff and Kim's (2013) term, and economics of promises as Pierre-Benoît Joly (2013) puts it: 'anticipations of positive futures' that justify the arguments developed and, more importantly, the investments in urban technologies. Observing these webinars enables us to see how rationalities and technologies trickle down or move up: how scales are connected.

Webinars perform the desire for technology-driven futures across the state while indirectly revealing the technological (in)capacities and resources of the state itself, in their absence. Although we did not directly analyse these moments, loss of video and/or sound, dropped connections, inability to un/mute voice and/or conform to online webinar protocols was a recurring feature of many state presentations. This wider context spoke to the challenges of a 'digitalising state' (Datta, 2023) where despite aspirations of ubiquitous virtuality, the faultlines of state resourcing and official capacity are laid bare for both stakeholders and citizens to observe.

In total we observed ten webinars (between 1 and 1.5 hours each) from April to December 2020, with the last one being with the World Economic Forum (WEF). Three webinars were on 'Leveraging Data and Technology in City's COVID Response- India', related directly to the pandemic during April to May 2020; two webinars were part of a series titled 'Tech Clinic'; three on a 'City Speaks' series; and two more on 'India-UK webinar for COVID-19 recovery', upholding collaborations and knowledge exchanges between the Indian Smart City Mission and global experts.

Webinars as political-technological performance of a crisis

In the webinars, crisis was a political-technological construct. It framed a state ideology where crisis is an acceleration towards a resilient future. As a professional consultant from a private technological company put it, the pandemic crisis has "fast-tracked strategic decisions which we otherwise would have taken a long time to take". He refers here to video-conferencing and sharing databases while others in the webinars mentioned the fast-tracking of the use of drones or face recognition.

These 'strategic decisions' referred to were invariably approaches in implementing smart technologies for state surveillance and monitoring (Figure 7.1). They were fast-tracked because they were mobilized in the context of a crisis. Zohar Sharon, chief knowledge officer of Tel Aviv municipality, noted during the WEF webinar: "This pandemic is a war". But

Figure 7.1: Screenshot of a webinar presentation of the Varanasi war room dashboard

the crisis was also contextual and historical, given the Israeli state's long-term preparation for biological attack. Thus, they were quick to deploy a GPS-based contact-tracing app called 'the shield'. This was a crucial moment, since the Indian officials who were in the same panel agreed that such long-term preparedness was an important aspect of dealing with crisis in the future.

The scope of using technology or interventions based on technological advancement were equally supported and highlighted by several officials from corporate sectors who were already part of smart city offices in different Indian cities. For example, Murat Sonmez, head of Centre for the Fourth Industrial Revolution (C4IR), WEF, speaking in 'Leveraging Data and Technology in City's COVID-19 Response' held on 29 May 2020 noted, "If we have the right data then by using AI and machine learning, we can improve the lives of citizens and not just cities". According to Sonmez, if people have wearable sensors that can track their vitals, 'then by using remote diagnostic, we can use drones to send medicines. This way, dependency on hospitals can be reduced or minimized'.

Such imaginations of crisis response are political-technological since they produce a technological response to dealing with a political problem of broken or absent welfare infrastructures. One of the recurring themes of these webinars, therefore, was COVtech solutions, which included several technologies of managing a pandemic – ranging from tracing COVID-19 patients, designating containment zones, and telemedicine, to a more systematic integrated system (property tax, smart waste management system, smooth traffic control, and so on). The webinars shared information across different smart cities on progress in repurposing ICCC's into COVID-19 war rooms. The CEO of Varanasi Smart City urged that if smart cities have yet not incorporated this, then they must act towards it urgently. Another official said that the most important thing is to create a common city data model, "which helps integrate the ICCC and the master plan".

Three articulated terms stood out in the political-technological construct of pandemic in the webinars: crisis, data, and resilience. *Crisis* evidently referred to the pandemic, but it was also 'contextual'. In other words, the pandemic was not considered as a state of exception. It was, as the SCM director repeated, 'the new normal'. And here it is difficult not to hear Naomi Klein's (2007) shock doctrine thesis: 'Crisis is an opportunity'. *Data* was thematized in different ways – as oxygen or oil – so the discourses pointed to the centrality and interoperability of data in the contemporary city. Data, the officials argued, was an accelerator to prepare for more efficient city management in the future. Finally, crisis is related to the last term in this articulated discourse: *resilience*. A computational response to crisis is not aiming to seek deep causes or solve the problem, or even to adapt to the crisis; rather, to build technological resilience for future urban development. This triad – data/crisis/resilience – is what Halpern and Mitchell (2023) argue as the core of the 'smartness mandate'.

In the following sections, we pick up the use of these three phrases in the webinars and war rooms of the Indian state's management of the pandemic. We argue that crisis is a political-technological construct that requires a politics of scaling up and down across various levels of the state, while data feeds the crisis through its geographic unevenness, thus producing a set of 'guarded invisibilities' (Furlong, 2020) within the war rooms. Resilience then emerges as something that is done to citizens rather than what citizens broker for themselves.

Scaling responses to the crisis

Webinars were simultaneously means for the orchestration of technopolitics in pandemic times and the stage where their logistics were made visible. However, one of the key features of this technopolitics was their scalar relations across different levels of the state. For example, 'learning' in these webinars was not just parallel – that is, between cities – but more significantly scaling up the pioneering technological advances of some cities to the entire country. As Cinnamon and Powell both argue in this volume (Chapters 10 and 4), data-driven urbanism is centrally about a politics of scale; the webinars orchestrated this scale through the performance of a technopolitics of crisis. In India, certainly, scalar technopolitics was part of the early development of the SCM when a new ruling party in 2014 promised to scale up the Gujarat model (a regional state seen as the pioneer in creating the first Indian smart city) to 100 smart cities across the entire nation (Datta, 2015).

To start with, the Indian Smart Cities Mission has been described by several scholars as a top-down vision (Hoelscher, 2016; Praharaj et al, 2018) designed with its headquarters in Delhi as a Latourian centre of calculation: a node in a socio-technical network where information, data, and knowledge are centralized and processed. Downstream, the centre dispatches norms, standards in terms of objectives, methods, budgets, and deadlines. Webinars regarding war rooms made this procedure very visible. Standardized procedures for reshaping ICCCs into war rooms came downstream as a handbook presented during one of the webinars. Entitled *Model war room for COVID-19* and published on 19 April 2020, it includes four sections covering in detail all aspects of the ICCC repurposing. The governance structure contains 22 functions for persons or teams, from the war room manager at the top to community collaboration at the bottom. The handbook also explains that a war room can be set up in the absence of an ICCC with a few desktops, LED screens, a printer, and an internet connection. The ICCC, as meta-infrastructure aiming to manage flows and circulations (Luque-Ayala and Marvin, 2020), serves as an adequate support for the centralization and management of data flows about various aspects of the pandemic (from GIS modelling of diffusion to the surveillance of lockdown compliance).

However, the webinars highlighted that scaling down from the federal state was simultaneous to the scaling up of war room models from a few pioneering cities. In one of the early webinars, the CEO of smart city and special officer of COVID-19 war room in Bengaluru municipality (BBMP) noted that the term 'war room' was coined by the Commissioner of BBMP. This was a significant point since Bengaluru Smart City (located in the regional state of Karnataka) was seen to be the pioneer in repurposing its ICCC into a war room to deal with the pandemic. The Bengaluru war room was seen as 'state of the art' with real-time integration of urban services, providing information on COVID-19 spread, conducting predictive modelling that led to Bengaluru becoming the first city to create a containment zone much before standard procedures on containment zones were taken up by the federal government. The Bengaluru war room also coded various apps for tracking people in these zones through aggressive contact tracing and using CCTV cameras and drones for monitoring movement. These apps also helped them manage the delivery of drugs, food, and rations, communications with resident welfare associations and NGOs to apprise themselves in real time of the changing situation on the ground.

Bengaluru's war room was therefore taken as a model war room for the rest of the country, and thus much of the early exchanges in the webinars focused on learning from Bengaluru. But Bengaluru's success in dealing with the COVID-19 crisis was not a coincidence as the Commissioner himself acknowledged. Bengaluru has a long history of being India's Silicon

Valley, or the IT capital of southeast Asia. The Commissioner attributed Bengaluru's success with COVID-19 war rooms to the critical mass of "tech-savvy people" who came to the city during the 1990s to leave a lasting legacy of material-technological transformations, including the continued attraction that the city holds for young IT entrepreneurs and new start-ups. He attributed the availability of talent and skills in the city, as well as its early success in installing one of the first Smart City ICCCs in Bengaluru that led to repurposing the first war room in the country in just 24 hours on 22 March 2020.

Uneven geographies of data

That the war room was an essential condition of the future was driven by a consensus among the ministry and urban local officials about extending war rooms to India's 100 smart cities. Representatives from the federal government suggested that the replicability factor for war rooms was being looked into, especially for ICCC's and digital technological solutions for all the cities. The webinars made it clear that the aim was ubiquitous surveillance; the way this would be made possible was through further extraction of data.

Data was a core component of the webinar discussions, either explicitly championed as 'oxygen or oil' or implicitly referred to as 'evidence' or 'facts'. The secretary to the government of India pointed out in one of the webinars that the role of ICCCs has been crucial during the pandemic, and "it worked as the nervous system of the body – collecting real-time data and other historical data". As he further noted on the expandability and interoperability of these systems, "Everything is completely data-driven and becomes so simple".

> 'Getting the information at one place. How do you integrate the whole thing (quarantine systems, hospital info, etc)? They work like the brain and nervous system. [...] Smart cities play an important role. [...] Every Smart City has a data officer, and we want to bring that to all 100 plus cities in India. We have a platform for data exchange. Ease of living is the goal of these data platforms. COVID-19 is only contextual. We must see further. We must have some kind of surveillance and discipline.'

Here data had a key role as evidence. Indeed, in the webinars, data became an arm of state governmentality whereby its role as a policy instrument was seen as a pathway to future resilience. As the home minister argued: "Let us take decisions that are based on evidence and data [...] Let us take our decisions on the base of solid data. Cities have been empowered to fight the pandemic with the help of data and technology together". The SCM director also noted that "slowing down the spread of COVID-19 is going to require, among other things, a heavy reliance on India's data infrastructures – providing real-time data readings for critical decision-making – and its Smart Cities Mission." Such opinions were common in other webinars as well. In these accounts, data was seen as disembodied matter that could generate value in its generalization and deemed objectivity.

However, as Graham et al (2015) argue, data is geographical. This became starkly evident during the pandemic as the spread, control, and management of the virus became an exercise in geographical knowledge of the state. This geographical knowledge had been accumulated over decades of state initiatives around e-government, whereby cities had been creating geographical databases, and under the SCM programme recently: digitizing land documents to increase revenue, matching property and tax records with GIS and Google Earth software. This was evident in the use of GIS data and the partnerships with Esri and Google which was critical to articulating a state approach to the pandemic crisis.

The webinars presented geographical data as a fundamental aspect of the war room. The IT expert in Varanasi Smart City believed that this pandemic was probably the best time to make use of GIS. In the webinar 'Leveraging GIS for City Operations' held on 25 May 2020, he narrated how the KICCC (Kashi Integrated Command and Control Centre) after being converted into a COVID-19 war room was utilizing its GIS resources to deal with "tough situations". Similarly, Scanpoint – another private technology partner – had provided mobile applications with features such as GPS, Bluetooth, and facial recognition to many municipalities. Representatives from Microsoft exposed the services provided by its company in COVID-19 war rooms, contact tracing for the government of Assam, including integration and cloud platforms, health and family welfare, apps like Cova in Punjab and Chhattisgarh, and Corona Tracker in Maharashtra. During the second wave of the pandemic in India, Microsoft facilitated services to build ICCCs and implemented 'power business intelligence' to develop dynamic features. These presentations positioned the private sector and particularly global IT companies such as Microsoft centrally within the state, not just in partnership but rather as key drivers shaping and directing the war against COVID-19 from the war rooms.

However, while the technology companies and state officials performed the rhetoric of objectivity embedded in COVID-19 data, embedded inequalities became evident from the assumptions in their geographic analysis. It was during one of the first tech webinars that a representative from a company called Infinite Analytics presented its new software to track and trace COVID-19 geographically. The visual and argumentative power of selected geographical data was here spectacularly staged to the participants of the webinars. Tracing began, the consultant said, from the monitoring of smartphones of those who attended what is now well known as a 'superspreader' event – the Tablighi Jamaat congregation, which contributed to the first wave of the pandemic in India.[1] Here is an excerpt of the consultant explaining this.

'We were able to track 929 participants in the Tablighi Jamaat using our platform. We were able to go back 15 days before they attended the Tablighi Jamaat, looked at where did they all come from, who all did they meet with, and during the event we were able to see what they were doing and where all did they go to. Not only that, but we were also able to find the first-degree contacts of those people as well [...] Essentially all the first-degree contacts that came in touch with the 929 attendees we were able to identify, and each of them we were able to give the authorities their ID, that is like a cookie on the phone as well as the carrier.'

This explanation by Infinite Analytics about surveillant mobility tracking using identifier for advertisers (IDFA) was given while webinar participants were shown 'night-time' animated images with moving yellow, red, or green dots representing the attendees of the event with zoom-in and zoom-out movements. The visual display of images is post-representational or 'post-cinematic' (Rose, 2022), using the potential of digital visualizations (animation, variation of viewpoints) as power to affect and impress. These representations resonate with what Louise Amoore notes as 'calculations of the algorithm appear to translate the observations of uncertain and contingent human life into something with the credibility of scientific judgement' (2009, 55). The concern for Tablighi Jamaat was also evident in some of the real-time comments that were received as the webinar continued and served to justify the state's approach to monitoring particular types of bodies in the war rooms. Here technology was a panopticon, and technology was also a biopolitical instrument of measuring the geography of the virus. Yet technology itself was biased, as Eubanks (2018) notes, in 'algorithms of oppression' whereby particular events and populations appeared to be selectively observed and tracked by the high-tech capacities of the phone and software. This emerges as the approach to tracing Tablighi Jamaat participants is in stark contrast to the second superspreader event, which contributed to a far more devastating second wave in India – the congregation of the 12-yearly Hindu Kumbh Mela, which was brought forward by a year to celebrate the perceived end of the pandemic in India.[2]

The uneven geographic surveillance constructed and perpetuated through the war eooms became evident particularly during the migrant exodus from cities that followed the first lockdown in April 2020. These migrants were particularly challenging to track and trace and monitor owing to their invisibility in official records and often their lack of access to mobile phones, which led to a humanitarian crisis of mobility, food, and shelter during this period. As Datta (2020, np) has argued, 'migrants are the survival infrastructures of Indian cities' and yet they were unaccounted for and largely unsupported by the state during this crisis. The webinar exchanges provided reassurances of safe return of migrants through the creation of migrant databases. As Sunil Kumar, the CEO of Smart City Bhagalpur, said:

> The city is expecting over 1 lakh [1 hundred thousand] migrant labour to return, and we are prepared to receive them, perform thermal scanning, quarantine them for 21 days and ensure proper food distribution. To tackle the future issue of employment for them, we are creating a database and will map them to jobs according to their skillsets. We are monitoring if anyone wants to start something of their own to encourage entrepreneurship, as well as attempting to employ them using MNREGA.
>
> (TimesTech, 2020)

These examples illustrate how government can extend the reach of technology beyond the pandemic enumerating community (in this case migrant) and 'mapping' them both geographically and socio-economically. Here, as Castells (2010) argues, – the 'power of flows takes precedence over the flows of power. Presence or absence in the network and the dynamics of each network vis-à-vis others are critical sources of domination and change in our society' (p 500). COVID-19 not only made migrants visible to the state, it also enabled the state to capture more data without necessary checks on privacy and consent.

The data politics operating in the above examples, enact a specific form of power that Isin and Ruppert (2020) call 'sensory power'. Drawing on Foucault, they write that 'personalised, miniaturised and distributed computing since the 1980s, and apps, devices and platforms especially since the 1990s, that facilitate tracking and tracing' (2020, p. 2), have enabled a new sensor-based form of governmentality. This does not upend previous power regimes; rather, it adds new modalities of power directed at particular social groups and marginalized populations. These modalities target clusters of people and relations, rather than bodies, populations, or territories.

But this sensory power is a utopian fantasy – a smokescreen that performs what is expected of governmentality. This power requires an elaborate performance in webinars through the visual simulation of data and its colonization beyond the technologies of the war rooms to how the data on all citizens (particularly migrant citizens and minority communities) is collected, stored, and controlled.

'Guarded invisibilities' of resilience

A central discourse in the webinars was on ensuring future state preparedness in case of emergencies, and resilience emerged as a key term repeatedly used by state officials. Referring to citizen participation as central to a lasting legacy of the war rooms, the SCM director noted:

> 'We should have predictive systems in slums to identify where the next epidemic is going to happen, where the next dengue case is going to happen [...] We need to reimagine this word called 'resilience' [...] these are the fundamentals of resilience.'

This focus on slums is not a coincidence. This follows what other scholars have argued as the seductions of 'broken data' – spaces where it is incredibly hard to connect the 'relationalities of data' (Mertia, 2020) emerging from bodies, geographies, and technologies. Slums have long represented a space of broken data for the state, which in the context of war rooms was seen to offer renewed potential of new harvests of information and knowledge.

While data was constructed as core to the long-term resilience of war rooms, it emerged in another webinar that its biggest challenge was the simultaneous unavailability of data on the ground. Ground data was deep data; it was gritty, complex, and temporal, and was hard to reach. So, in many instances, the support of academic and non-governmental organizations became crucial to 'fix' this data scarcity by addressing its brokenness, disjointedness, and thinness in hard-to-reach geographies. Municipal health officials, activists, NGOs, and civil society groups collected this data on behalf of the state, enriching the models and real-time visualizations that we then see in the war rooms. And it was the few presentations by CSOs and internet activists in later webinars who were supportive of collecting deep data for analysis that also raised concerns about how technologies in the war rooms were implemented without clear legal or constitutional framework.

In this context it became clear that data resilience was reliant on testing and triangulating data on the ground by civil society organizations and activists who were directly engaged with social groups who had been excluded from state and private-sector databases. These organizations were also custodians of deep data, which was necessary to extract value from the big data accumulated through tracking devices in the war room. As Sarah Barns (2018, 11) has noted, while the dashboards in the war rooms 'showcased' data, 'like an iceberg, perhaps slowly melting, that which is visualized and revealed by urban data platforms may not, in fact, be the whole story'.

This becomes clear during the Q&A sessions of the webinars, when several questions from the audience were raised about the fidelity of collected data,

Figure 7.2: (a, b, c) Screenshots of webinar Q&A sessions

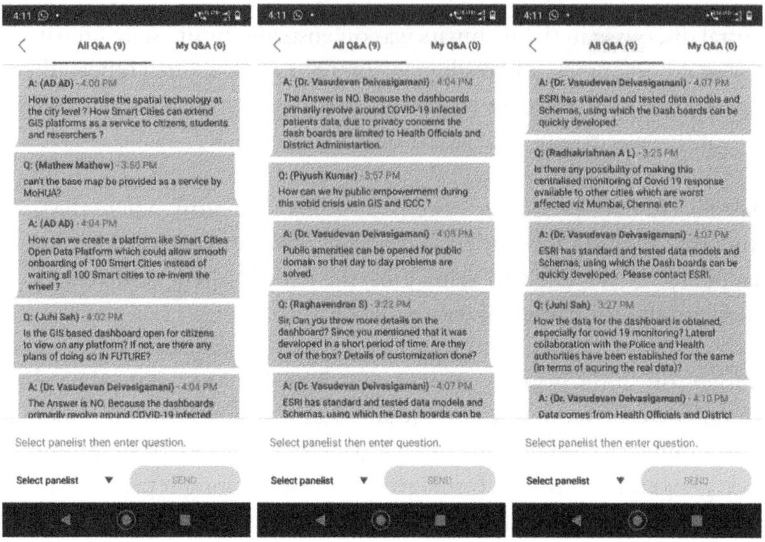

particularly in relation to what would constitute 'real data' on ground, how this data is analysed, stored, and potentially shared. Often questions from the audience would push towards wider use of data for public good, which was seen by state officials as conflicting with data privacy (Figures 7.2 a, b and c). At the same time, questions about data privacy from the audience would lead to reassurances from the consultants and state officials that personal information was masked and privacy was strictly maintained through access control.

> Q: How the data for the dashboard is obtained, especially for covid 19 monitoring? Lateral collaboration with the Police and Health authorities have been established for the same (in terms of acquring [sic] the real data)?
>
> A: Data comes from Health Officials and District Administration every night in conventional forms which is then collated, entered in the Excel tables which is then used to update GIS servers which automatically updates the dashboard.

The IT expert (Varanasi Smart City) mentioned that certain individuals and CSOs are worried about data security.

> 'There has been some objection to the privacy matter, so I just masked it, but the people, the health authorities and other district authorities are able to see the patient, their location, the vital statistics as data.'

The question of data resilience seemed then to be fraught with challenges. While data harvesting was seen as a pathway to future resilience against crises, there was serious data scarcity in marginal geographical spaces. And while there was data scarcity, deeper data extraction was deployed through community organizations. Yet as the state became the custodian of data extracted during 'war efforts' in the ICCCs, it was reluctant to democratize and open up data for use outside its war rooms. The federal state on the other hand, relied on the decentralized labours of various municipal administrations in order to bolster a seamless vision of calculability and predictability of crisis. And yet women, migrant workers, transgender communities, and service workers who were the most affected were invisible in the spatial analysis of tracking and tracing devices. So while the state engages in what Barns (2018) calls a 'data showcase' in the COVID-19 war rooms, it also shows how public access to this data is restricted by notions of security, privacy, and disconnectedness from local scales.

Conclusion

The webinars we observed and the war rooms that are represented therein are the new echo chambers of the state. While the state may indeed appear to have become non-hierarchical in the webinars and war rooms, power continues to be vested in the centralized state that can withhold both resources and information from other levels of the state, even as it is often reliant on the urban level for learning and scaling up knowledge and a way to do things. The data that flow into the war rooms are colonized and ringfenced, and thereby create digital enclosures that are highly prescriptive of who or what is in crisis, and who and how they should become resilient.

The significance of the webinars as the primary mode of technopolitics during the pandemic produced a discourse of their legacy as the 'new normal'. This is highlighted by the SCM director in the tech clinic webinar:

'What we were not doing before COVID is exactly what we are doing now. This is a fantastic way of connecting with the world, which should become a new normal post the COVID crisis also.'

Important work has been done especially by authors of other chapters in this volume on dashboards and platforms (Rob Kitchin, Federico Caprotti, Nancy Odendaal). They all suggest how technopolitics demands both rationalities (socio-technical imaginaries) and technologies (war rooms and webinars) through the smartness mandate. We have argued that in India the smartness mandate is much fuzzier and more dispersed across scales and spaces of the state and para-state, while it is also shaped by the genealogy of technology development as a tool of modernity and development since the postcolonial era. In India, smartness comes with its limits because of acute data scarcity and infrastructural disconnectedness, which means that the rationalities of the war rooms remain more performative than instrumental. The glitzy presentations and floating dots of COVID-19 infections on the screens of the consultants bolster the sensory power of war rooms while simultaneously articulating the geographies of data as a terrain of state power.

The war rooms and webinars highlight the broader issues associated with normalizing crisis as an opportunity. This rhetoric can produce crisis itself as the articulation of a new kind of sensory power of the state through these vision machines. Yet the context of this development varies vastly from the technopolitics of smart cities mandates. India both allows and disallows pathways to data extraction and availability, creating both 'guarded invisibilities' (Furlong, 2020) and scaled visibilities from the city to the federal state. In this context, learning across cities to deal with crises appears to be far more non-hierarchical and decentralized than other smart cities globally, yet the federal state also remains the main orchestrator of data politics, determining what data should be collected, how that should address the crises, and who is to become resilient as a result of these actions.

Acknowledgements

We would like to acknowledge the generous support of the Swiss National Science Foundation (SNSF). This chapter is one of the results of the SNSF-funded research project Smart Cities: 'Provincializing' the global urban age in India and South Africa (SNSF grant: 10001AM_173332). We would also like to acknowledge the collective work of our researchers Arunima Ghoshal, Arya Thomas, Anwesha Aditi, and Yogesh Mishra in assisting with the work that has made this chapter possible.

Notes

[1] A Tablighi Jamaat religious congregation that took place in Delhi's Nizamuddin Markaz Mosque in early March 2020 was called a COVID-19 superspreader event, with more than 4,000 confirmed cases and at least 27 deaths linked to the event reported across the country. Over 9,000 missionaries may have attended the congregation, with the majority being from various states of India, and 960 attendees from 40 foreign countries. On 18 April 2020, 4,291 confirmed cases of COVID-19 linked to this event by the Union Health Ministry represented a third of all the confirmed cases of India. Around 40,000 people, including Tablighi Jamaat attendees and their contacts, were quarantined across the country.

[2] Millions attended the Kumbh Mela as India's second COVID wave took off. There were 1.2 million estimated daily visitors to Haridwar during the festival. 2,000 people tested positive for COVID-19 in Haridwar, 10–14 April. During the Mela there was an 1,800% increase in COVID cases in Uttarakhand state.

References

Amoore, L. (2009) 'Algorithmic war: Everyday geographies of the war on terror', *Antipode*, 41: 49–69. https://doi.org/10.1111/j.1467-8330.2008.00655.x

Barns, S. (2018) 'Smart cities and urban data platforms: Designing interfaces for smart governance', *City, Culture and Society*, 12: 5–12. https://doi.org/10.1016/j.ccs.2017.09.006

Caprotti, F. (2019) 'Spaces of visibility in the smart city: Flagship urban spaces and the smart urban imaginary', *Urban Studies*, 56(12): 2465–79.

Castells, M. (2010) *The Rise of the Network Society, The Information Age: Economy, Society and Culture, vol 1* (2nd edn), Oxford: Blackwell.

Chang, I.-C.C., Jou, S.-C. and Chung, M.-K. (2021) 'Provincialising smart urbanism in Taipei: The smart city as a strategy for urban regime transition', *Urban Studies*, 58(3): 559–80.

Datta, A. (2015) 'New urban utopias of postcolonial India: 'Entrepreneurial urbanization'in Dholera smart city, Gujarat', *Dialogues in Human Geography*, 5(1): 3–22.

Datta, A. (2018) 'The digital turn in postcolonial urbanism: Smart citizenship in the making of India's 100 smart cities', *Transactions of the Institute of British Geographers*, 43(3): 405–19.

Datta, A. (2019) 'Postcolonial urban futures: Imagining and governing India's smart urban age', *Environment and Planning D: Society and Space*, 37(3): 393–410.

Datta, A. (2020) 'Survival infrastructures under COVID-19', Geography Directions blog, [online] 13 May, available from: https://blog.geographydirections.com/2020/05/13/survival-infrastructures-under-covid-19/

Datta, A. (2023) 'The digitalising state: Governing digitalisation-as-urbanisation in the Global South', *Progress in Human Geography*, 47(1): 141–59.

Datta, A., Aditi, A., Ghoshal, A., Thomas, A. and Mishra, Y. (2021) 'Apps, maps and war rooms: On the modes of existence of "COVtech" in India', *Urban Geography*, 42(3): 382–90.

Deane, C. (2015) 'The Control Room: A Media Archaeology', *Culture Machine* 16, [online], available from: https://culturemachine.net/vol-16-drone-cultures/the-control-room/

Dunleavy, P., Margetts, H., Bastow, S. and Tinkler, J. (2006) *Digital Era Governance: IT Corporations, the State, and e-Government*, Oxford: Oxford University Press.

Eubanks, V. (2018) *Automating Inequality: How High-Tech Tools Profile, Police, and Punish the Poor*, New York: St Martin's Press.

Furlong, K. (2020) 'Geographies of infrastructure II: Concrete, cloud and layered (in)visibilities', *Progress in Human Geography*, 45(1): 190–8. https://doi.org/10.1177/0309132520923098

Gabrys, J. (2014) 'Programming environments: Environmentality and citizen sensing in the smart city', *Environment and Planning D: Society and Space*, 32(1): 30–48.

Goodspeed, R. (2015) 'Smart cities: Moving beyond urban cybernetics to tackle wicked problems', *Cambridge Journal of Regions, Economy and Society*, 8(1): 79–92.

Graham, M., De Sabbata, S. and Zook, M.A. (2015) 'Towards a study of information geographies: (Im)mutable augmentations and a mapping of the geographies of information', *Geo: Geography and Environment*, 2(1): 88–105. https://doi.org/10.1002/geo2.8

Haklay, M. (2013) 'Citizen science and volunteered geographic information: Overview and typology of participation', in D. Sui, S. Elwood and M. Goodchild (eds) *Crowdsourcing geographic knowledge: Volunteered Geographic Information (VGI) in Theory and Practice*, Dordrecht: Springer, pp 105–22.

Halpern, O. and Mitchell, R. (2023) *The Smartness Mandate*. Cambridge, MA: MIT Press.

Halpern, O., Mitchell, R. and Geoghegan, B.D. (2017) 'The smartness mandate: Notes toward a critique', *Grey Room*, 68: 106–129.

Haraway, D. (1988) 'Situated knowledges: The science question in feminism and the privilege of partial perspective, *Feminist Studies*, 14(3): 575–99. https://doi.org/10.2307/3178066

Hecht, G. (2009) *The Radiance of France: Nuclear Power and National Identity after World War II* (2nd edn), Cambridge, MA: MIT Press.

Hoelscher, K. (2016) 'The evolution of the smart cities agenda in India', *International Area Studies Review*, 19(1): 28–44.

Isin, E. and Ruppert, E. (2015) *Being Digital Citizens*, London: Rowman & Littlefield.

Isin, E. and Ruppert, E. (2020) 'The birth of sensory power: How a pandemic made it visible?', *Big Data & Society*, 7(2).

Jasanoff, S. and Kim, S.-H. (2013) 'Sociotechnical imaginaries and national energy policies', *Science as Culture*, 22(2): 189–96.

Joly, P.-B. (2013) 'À propos de l'économie des promesses techno-scientifiques', in *La Recherche et l'Innovation en France*, FutuRIS 2013, Paris: Odile Jacob, 231–55.

Kitchin, R., Maalsen, S. and McArdle, G. (2016) 'The praxis and politics of building urban dashboards', *Geoforum*, 77: 93–101.

Klein, N. (2007) *The Shock Doctrine: The Rise of Disaster Capitalism*, London: Macmillan.

Latour, B. (1987) *Science in Action: How to Follow Scientists and Engineers through Society*, Cambridge, MA: Harvard University Press.

Luque-Ayala, A. and Marvin, S. (2020) *Urban Operating Systems: Producing the Computational City*, Cambridge, MA: MIT Press.

Mattern, S. (2015) 'Mission Control: A History of the Urban Dashboard', *Places Journal*, [online] March, available from: https://placesjournal.org/article/mission-control-a-history-of-the-urban-dashboard/

Meijer, A. (2018) 'Datapolis: A public governance perspective on "smart cities"', *Perspectives on Public Management and Governance*, 1(3): 195–206. https://doi.org/10.1093/ppmgov/gvx017

Mertia, S. (ed) (2020) *Lives of Data: Essays on Computational Cultures from India*, Theory on Demand 39, Amsterdam: Institute of Network Cultures.

Morozov, E. (2013) *To Save Everything, Click Here: The Folly of Technological Solutionism*, New York: Public Affairs.

Praharaj, S. (2020) 'Development challenges for big data command and control centres for smart cities in India', in N. Bilimoria (ed) *Data-Driven Multivalence in the Built Environment*, S.M.A.R.T. Environments, Cham: Springer, pp 75–90.

Praharaj, S., Han, J.H. and Hawken, S. (2018) 'Urban innovation through policy integration: Critical perspectives from 100 Smart Cities Mission in India', *City, Culture and Society*, 12: 35–43. https://doi.org/10.1016/j.ccs.2017.06.004

Rose, G. (2022) 'Introduction: Seeing the city digitally', in G. Rose (ed) *Seeing the City Digitally: Processing Urban Space and Time*, Amsterdam: Amsterdam University Press, pp 9–33.

Söderström, O. (2021) 'The three modes of existence of the pandemic smart city', *Urban Geography*, 42(3): 399–407.

Sukumar, A.M. (2019) *Midnight's Machines: A Political History of Technology in India*, New Delhi: Penguin.

TimesTech (2020) 'How smart cities are talking the COVID-19 pandemic', *TimesTech*, [online] 30 May, available from: https://timestech.in/how-smart-cities-are-tackling-the-covid-19-pandemic/

Zook, M. and Spangler, I. (2023) 'A crisis of data? Transparency practices and infrastructures of value in data broker platforms', *Annals of the American Association of Geographers*, 113(1): 110–28. https://doi.org/10.1080/24694452.2022.2071201

8

The Smartmentality of Urban Data Politics: Evidence from Two Chinese Cities

Ying Xu, Federico Caprotti, and Shiuh-Shen Chien

Introduction: Urban data politics and the Chinese smart/platform city

In contemporary cities in China, and further afield, there are various examples of experimental urbanism that link the smart city and sustainable futures (Caprotti, 2019b). Over the past few decades, multiple urban paradigms and themes have been developed and applied to try to experiment with new formats for the future city. Examples of this, stretching back into the 20th century, are eco-cities, low-carbon cities, and more recently (especially in the 2010s), smart cities. More broadly, the two decades between 2000 and 2020 have seen a shift in focus from an interest in green and sustainable urbanism (of which the eco-city, eco-towns, and eco-neighbourhoods were prominent examples) to the development of strategic urban visions predicated on the use of smart technologies, data analytics, and sensor networks (Joss et al, 2022). More recently still, it has been argued that the smart city has been superseded by the growing trend known as platform urbanism (Caprotti et al, 2022), within the broader context of digital capitalism (Törnberg, 2023), in which digital platforms become pre-eminent interfaces that enable data intermediation within and across cities, as seen by the rise of platforms such as Uber, Airbnb, and Didi Chuxing, or by the development and implementation of urban governance platforms such as Hangzhou's City Brain (Caprotti and Liu, 2020a; 2020b).

In China, as elsewhere, state-led urban development programmes are constantly seeking to revitalize and innovate human settlements and the urban landscape (Shen, 2020). Since the early 2010s, China's smart cities landscape took shape among discourses of technological innovation (from sensor systems to artificial intelligence (AI) and Big Data) and economic restructuring (focused on Chinese integration into, and leadership of, the global digital economy) (Caprotti and Liu, 2020b; Cugurullo, 2021; Duffie and Economy, 2022). In part, these discourses are standard replications of the sorts of discourses, narratives, and imperatives repeated throughout smart city strategies and projects worldwide. These narratives include notions like efficiency, speed, and the promise of the frictionless digital city, and include a prominent turn to the market and technology corporations in the continued development of neoliberal smart urban governance (Grossi and Pianezzi, 2017). However, this is not necessarily the case in the Chinese context, where strong central state steering in terms of economic and urban development themes, and responsive provincial and municipal governments, all play a leading part in developing and translating urban development priorities and models, from the eco-city to the smart city and beyond.

The Chinese Social Credit System (SCS) is an example of the evolution of smart urbanism into platform urbanism. It is a system based on the intermediary function of a social credit platform (usually accessed by urban residents through smartphones) through which urban government can interface with residents, and through which residents' actions, consumption, and other behaviours become knowable and recorded. The core promise of the SCS is in the production of social harmony through the performance of citizenship as defined by authorities through the social credit platform, its rating system (Dai, 2020), and its myriad incentives and penalties (Zhang, 2020). The SCS is also an example of the spatially variegated character of platform urbanism, since the national vision for social credit is being developed and operationalized across multiple but separate (for now) municipal-scale SCSes.

Based on the following municipal cases in China, two primary arguments become clear. The SCSes are taken as innovative and increasingly ubiquitous governing strategies and platforms around not only the smart city but also focused on reshaping and recasting citizenship via a key characteristic of credit systems, namely the relational performance that occurs through smartphones on the one hand, and through digital, algorithmic governance on the other. Moreover, when smartmentality is operationalized in the case of the SCS, the state displays a significant and highly pervasive agency in governing *through* the reshaping of (digital) urban citizenship and related, embedded data politics. This goes beyond oft-repeated and explored links, in much of the urban literature, between the state, platform, and algorithmic governance, and neoliberalism. The development of a platformized form of smartmentality in Chinese cities is, we argue, an example of a specific, geographically contingent evolution in the imbrication between the state and citizens. In this evolution, the private sector necessarily enables the technique of smartmentality, while the driving agency, control, and regulation of the 'system' remains in the hands of the state.

Therefore, from its beginnings with the national social credit pilot programme in 2014, the SCS project was based on notions of digitally and platform-enhanced urbanism, governance, and economic efficiency, much like more globalized notions of the smart and platform city. However, what sets the SCS aside is its focus on deep engagement with urban residents through promoting and attempting to shape citizens. At the core of this endeavour is the notion of what we term citizenship-as-performance, whereby digital platforms enable a form of digital urban governance that promotes the performance of 'good' citizenship tasks, and the alignment of urban life with state-steered priorities. This can be understood, conceptually, through the application of the concept of smartmentality to the SCS, and it is to this that we now turn.

Smartmentality, the Social Credit System and the Chinese city

We base our understanding of the SCS on an evolution of Foucault's concept of governmentality. This has been used in critical analysis on urban societal issues, such as moral discourses, public opinion and policies, and scientific knowledge (Burchell et al, 1991; Crampton and Elden, 2007). Governmentality generally refers to organized mechanisms and practices through which governments (and other actors) influence groups' and individuals' conduct, including through the production of specific cultural processes (Rose, 1999). The historically and geographically contingent process of governmentality involves controlling, inciting, enabling, promoting, or suppressing human actions. More specifically, those actions include conduct, identities, and other aspects of human life.

A growing body of research has established a conceptual and empirical link between work on governmentality and research on smart cities and smart citizenship (Gabrys, 2014; Vanolo, 2014; Rodrigues et al, 2022). This has resulted in the development of a focus on 'smartmentality', as an emerging normative disciplinary mechanism operationalized in smart cities and through smart city policies and politics (Vanolo, 2014). A corollary of the establishment of smartmentality through the smart turn in urban development has been a renewed focus on digital citizenship and its rearticulation through urban data practices (such as monitoring, distribution, and feedback around datafication processes). There is a continuing need for analysis of smartmentality-focused interventions in various geographic contexts (Scannell, 2015; Kitchin et al, 2017).

At the same time, there has been little focus on the ways in which smartmentality is operationalized in the Chinese context. As Zhang et al

(2022) have noted, much research on smartmentality focuses on non-Eastern contexts, from the UK, to South Africa, to cities in the USA and Europe, and links smart urban development with questions around iterations of neoliberalism. Zhang et al (2022) argue that while the majority of the current literature focuses on the link between smart cities and the market under the umbrella of neoliberalism, in China smartmentality is defined by a dynamic of power that sees state steering as the prime mover in attempts to discipline and influence citizenship and behaviour. This change of focus is key, in part because it recognizes the importance of engaging with urban development trends in a Global East context (Müller and Trubina, 2020) and with the production of urban knowledge in a more decentred manner. Additionally, it helps to decouple the concept of smartmentality from neoliberal, capitalist logic and therefore broadens the scope of research on its application.

In addition to contributing to linking smartmentality with the Chinese context, we also aim to show how smartmentality works not just in standard smart city contexts, but also in the emerging landscape of platform urbanism. The Chinese SCS is a prime example of the growing trend of the use of platform-based digital governance interfaces in ways that connect, in depth, with the manner in which smartmentality works to influence, discipline, and construct and perform behaviour. With smart urbanism, smartmentality tended to be expressed in specific urban contexts and linked to policies, technologies, and initiatives at the city scale. In contrast, the SCS shows an overarching logic that is set at a national scale, and that leverages policies and strategic priorities that exist over and beyond the city or even provincial scale. While different Chinese cities have developed their own SCSes, these are all in the broader context of the national SCS programme, which is predicated on state steering.

It is also clear that smartmentality is extending beyond the specific, stated remit of SCSes. For instance, in the context of the COVID pandemic (measures and restrictions related to this are still in place in China at the time of writing), the digital COVID 'health code' used across Chinese cities is an example of the creeping use of digital platforms for purposes of social control. On the one hand, digital COVID health codes aim to enable rapid responses to pandemic situations through providing a colour-coded rating of an individual's health risk. This rating, in turn, determines the individual's ability to access transport and other facilities, and can lead to situations such as quarantine. However, there has been recent media coverage in the *South China Morning Post* of those who claim that the COVID code was not used for public health reasons, but for social control: individuals have reported their COVID codes turning red (and therefore triggering mandatory quarantine) following their participation in demonstrations in Zhengzhou in May and June 2022 that protested the freezing of deposits of a bank in Henan province (Zhang, Zhang and Yang, 2022). The BBC reported that while citizens were subsequently told that their health codes would turn green (and they would be released from quarantine) after two negative COVID tests had been registered (following the code turning red), police notified those taking part in the protests that their codes would turn green if they left Henan (Wong, 2022). This links to the notion of crisis discussed below, where the boundaries and potential applications of SCSes, and their ramifications, are increasingly blurry and fluid, and open to manipulation for a range of aims not directly related to the objectives for which these platforms were designed. There is, therefore, an unfolding potential landscape of the use of SCSes in a 'lateral' sense: to respond to real or constructed crises *before* they arise, or *during* their development. The key guiding aim at the heart of the SCS programme – enabling social harmony and trust in society – is thus used to justify a definition of social harmony based on the activities that need to be excluded from the public (urban) sphere in order for this vision of harmony to be achieved and, more importantly, maintained.

A final point linking smart and platform urbanism with smartmentality in Chinese cities is linked to the notion of citizenship. Citizenship was not centrally considered in much of the literature on smart cities with a non-Chinese focus: research tends to underline themes such as participation, inclusion, social sustainability, the digital divide, and the like. Not until recent work on citizen–focused smart urbanism was the dominant (largely neoliberal) logic examined, and the (digital) rights to the (smart) city brought into a dispute (Cardullo and Kitchin, 2019; Kitchin et al, 2019; Cardullo, 2020). In many ways, smart cities literature has tended to be largely silent on the link between citizenship and smartmentality. In the Chinese city, there is a clear link between urban citizenship and smart and platform urbanism as enshrined in the SCS. The use of digital platforms such as SCSes, stretching across multiple social, economic, and other domains, in an effort to incentivize the performance of specific notions and characteristics of what is a 'good citizen', raises questions about the future direction(s) of the digital, smart, and platform-mediated city. In some ways, platform urbanism as seen in the SCS has potential repercussions in terms not only of urban citizens' agency within the city, but the city proper (as performed by authorities, individuals, corporations, and others) and its agency on citizens. Questions also arise around the extent to which smartmentality fosters a log of inclusion/exclusion, whereby citizenship (and its duties and rights) is experienced and acted on increasingly and solely through digital platforms that intermediate relationships between individuals and the state.

China's Social Credit System

It is useful, here, to provide a broad outline of the development and current trajectory of the SCS. At the time of writing, the SCS exists as a collection of municipal and private sector-run SCSes. These are largely unconnected one to the other, and have different characteristics. Private sector SCSes, such as Ant Financial's Sesame Credit, are used nationwide and are focused on trustworthiness in the realms of consumption and financial transactions. Municipal SCSes, on the other hand, are led and governed by city authorities, focused on urban services and governance, and usually limited in their use to specific spatial jurisdictional boundaries.

The diverse nature of municipal and private sector SCSes in existence today sits within a broader context of state-steered visions and strategies for developing a national approach to the SCS. As a high-profile national social project, there exists an overarching state-sponsored vision regarding the construction of the SCS. Its ambition is to provide a digital system which can be applied throughout China, as expressed in the State Council's 2014 *Outline of the Social Credit System Construction Plan (2014–2020)* (hereafter OSP). A pilot programme was carried out in ten secondary-tier (Weihai, Suzhou, Hangzhou, Yiwu, Wenzhou, Xiamen, Chengdu) or smaller-scale cities (Rongcheng, Weifang, Suqian). However, there had been several separate SCS experiments conducted by other cities since at least the early 2010s (Creemers, 2018).

The national SCS project has four major aims and targets, enshrined in the OSP: codes, incentives, societal governance, and the market mechanism. The first (codes) is to promote the construction of 'credit standards'; that is, the laws, regulations, and policy documents for the SCS. The second is incentives for trustworthiness, and penalties for lack of it. The third is the aim of building a new type of social governance mechanism based on credit. The underlying principle is that those whose behaviour is in line with the standards expressed through the SCS experience the system as a silent background, which sometimes benefits them with incentives; while those whose behaviour is not in line with SCS standards experience the SCS as an overarching and all-seeing presence through which 'negative' behaviour is increasingly weighted with public and private consequences. The fourth is to cultivate the credit service market. Under the government's guidance, the market mechanism is to be harnessed in order to enrol private sector actors in the co-production of trustworthy citizens, mostly through a focus on trust in transactions, payments, and consumption-focused behaviours. Indeed, much of the focus of the social credit market in municipal SCSes is on accumulated credit to be used in exchange for goods and/or services.

In the context of the expansion of everyday financialization globally, our SCS analysis highlights the ways in which people are involved in the datafication of behaviour, the commodification of consumption and other behaviour into tangible credit, and the (positive and negative) consequences of the deployment of algorithmic governance in providing incentives and penalties in this context. Moreover, the credit-based financialization approach evident in SCSes heralds a new urban and state landscape in which algorithms and calculative practices can (aim to) estimate and manage individuals (and their acts), as well as governing whole urban communities through the techniques, systems, and overarching (and digitally embedded) logics found within these systems. At a more general level, the citizenship and governance logics at the heart of SCSes are centrally concerned with the moral ordering of individual and group behaviour, through mechanisms such as ranking, sorting, rendering public, incentivizing and penalizing, enabling and inhibiting specific activities, behaviours, and possibilities.

At the core of the SCS endeavour is the ideal of deploying a ubiquitous web of platform-based technical systems through which behaviour can be sensed and ranked, and which can be used to steer individual and group actions and behaviour in 'desirable' directions. Central to this is the notion of social harmony enabled through a system that renders behaviours (selectively) transparent. Thus, while the algorithmic logics and technical make-up of SCSes may remain clouded or opaque (Fields et al, 2020), transparency is one of the key features of attempts to make citizens knowable and gradable by the state (Caprotti, 2019a; 2019c). Social harmony, then, is achieved (through the SCS) by leaning into the key characteristics of digital platform systems, such as automation, artificiality, self-determination, and existence separate from values and ideas that are external to the technical system (Ellul, 1962). Our argument, here, is that the development and deployment of the SCS are on the one hand based on the ideal of social harmony, but that this vision – of a harmonious polity and cohesive cities based on digitally enabled transparency – is predicated on avoidance of *disharmony*, and specifically on avoidance of the complex, messy, and potentially destabilizing interactions that are made possible by the continuing digital development of society. In this sense, then, SCSes can be seen as attempts to avoid looming, emerging, real, or imagined crises that may (or may not) spill out onto city streets and become articulated in digital spaces and through digital media. The disciplining role of SCSes then becomes apparent, as mechanisms which enable stability by staying one step ahead of disorder.

The local adoption and operation: municipal SCSes in two Chinese cities

In this section, we provide a brief outline of the development and timelines for SCS initiatives in Hangzhou and Tianjin, before analysing key aspects related to smartmentality in the next section. Both cities developed municipal-scale SCSes: Hangzhou was an early mover in developing a prototype municipal SCS since the 2000s. In addition, Hangzhou is one of China's leading digital cities, as seen in its roll-out of the City Brain urban governance system (Caprotti and Liu, 2020b; Zhao and Zou, 2021), and indeed in its roll-out of a municipal SCS programme in 2018. Tianjin, on the other hand, was a prominent developer of eco-urbanism through its flagship Sino-Singapore Tianjin Eco-City project, which was initiated in the late 2000s (Xu et al, 2022). Its SCS programme has taken on key facets of the characteristics of eco-urbanism.

Hangzhou

In Hangzhou, an early attempt to develop an SCS was the initial iteration of Credit Hangzhou (信用杭州), which began in 2002 (Credit China, 2018). Hangzhou was the first sub-provincial city that developed a credit plan (信用规划) since the 2006–2011 national 11th Five-Year-Plan for social and economic development. Since 2008, Hangzhou has operated a credit platform (信用平台) according to three major dimensions – urban governance, urban services, and informational public welfare. Meanwhile, in 2014 Hangzhou-based technology corporation Alibaba kicked off the first private sector SCS trials of Sesame Credit. Sesame Credit scores are based on credit history, behavioural trends, ability to honour financial agreements (for example, bills), personal information, and social relationships. Incentives available to Sesame Credit high-scoring individuals include reductions or waivers of deposits for a range of services, access to fast lanes at airports, and discounts (Figure 8.1).

Up to 2017, Hangzhou's credit platform was defined as a national demonstration project, connecting all municipal departments and main public service institutes. It incorporated 366 million pieces of credit information (across 338 categories comprising 2,947 items), covering a population of 29 million (including permanent residents and migrants to Hangzhou). In November 2018, the Hangzhou municipal SCS was officially launched and branded as Qianjiang Credit, with the intention of 'guiding citizens to be honest and good, and promoting the core values of socialism.' The SCS

Figure 8.1: The deposit-free renting (of mobile phone power banks) and 'Credit-easy+' services in Hangzhou

Source: photos taken by the authors.

relies on bringing together data from multiple areas of city life, including government affairs, economy, justice, life, public welfare, and other fields. The resulting database relies on the Hangzhou Public Credit Information Platform, the Hangzhou Government Affairs Data Resource Sharing Platform and the user data accumulated by the Hangzhou Citizen Card corporation over more than a decade, combined with the data from Zhejiang Provincial Development and Reform Commission and the Provincial Credit Centre. The SCS assesses a citizen through five key dimensions: demographic information, compliance with laws and regulations, livelihood, commercial integrity and pro-social behaviour. At the time of writing, 4.73 million citizens have joined the SCS and voluntarily authorized a Qianjiang credit score. In brief, Hangzhou's Qiangjiang Credit SCS is the result of a long-term process starting with initial SCS visions, through to collaboration with private sector Sesame Credit, to the municipal government-led Qianjiang Credit system.

Tianjin: from 'recycling credit' to municipal 'HaiRiver credit'

As a city recently enrolled in the national SCS scheme, Tianjin piloted a recycling credit programme since the early 2010s in the Sino-Singapore Tianjin Eco-City, which has attracted attention as an example of the eco-urban development paradigm (Caprotti, 2014; Pow, 2017). Since its beginnings in 2014, the Recycle Credit program in the eco-city has evolved from a standalone aim of incentivizing the transition towards a zero-waste city, and has become part of the eco-city's transition towards smart urbanism. Primarily concentrating on the green dimension, the SCS's main focus remains on enabling and promoting recycling, with the basic logic of the accumulation of credits through recycling, and the ability to exchange credits for goods.

Tianjin eco-city's SCS is based on three key elements. The first one is the waste and recycling facilities located in 19 urban communities. Regarding the spatial location, these facilities are all set within 150 meters' (or 2 minutes') walking distance of residential accommodation (Jinwanbao, 2021). Recycling different types of waste generates credits that can be awarded to residents' accounts on a smart information-management platform. For example, plastic is worth 80 credits per kilogram, and glass is worth 5 credits per kilogram when recycled. The second characteristic is the credit exchange shops, based in community centres. Here, shampoo, tissues, or other housewares can be exchanged at the rate of 100 credits for CNY 1. According to staff interviewed in the shop, an average of 3,000–5,000 credits a day (equal to CNY 30–50) have been awarded to residents. Third, the SCS features a smart platform and mobile app for residents. Data on disposal amounts and garbage types are collected and integrated automatically through the platform. With the mobile app, residents can access recycling facilities and check their recycling records, credit rules, and location of nearby waste facilities (Figure 8.2).

Figure 8.2: Main elements of the recycling credits programme in Tianjin eco-city

Source: photos taken by the authors.

Apart from the eco-city recycling credits programme, since 2021 the wider Tianjin municipality has implemented a social credit policy. Binhai New District, near Tianjin city, was selected as a national pilot city for SCS development. In the roadshow held by the municipal Development and Reform Commission and Public Credit Centre, the culture of integrity and selected moral criteria were communicated. This was also communicated to communities via slogans such as 'Integrity has a price', 'It is beneficial to be credible', and 'Credit is building civilization together'. More broadly, at the time of writing, a comprehensive municipal SCS, known as HaiRiver Credit, is being launched in Tianjin.

While Hangzhou and Tianjin's SCSes are focused on specific urban areas, there are increasingly interregional aspects to these SCSes. This is in line with the SCS being contextualized more within the spatially flexible remit of platform urbanism, rather than the more city-scale smart city project dimension. Tianjin and Hangzhou, for example, have agreed to recognize each other's credits. On 9 August 2019, Hangzhou, Nanjing, Wuhan, Suzhou, and Zhengzhou signed a framework agreement on individual citizen credit and joint incentives (Tianjin was included later), to achieve mutual recognition of cross-regional credit points and interoperability of application scenarios. Under this collaboration framework, Hangzhou citizens can enjoy benefits obtained through the city's Qianjiang credit, by using a Suzhou SCS card. In this small example, we are seeing the start of a trend toward intercity and interregional sharing, collaboration, and application of social credit and associated data.

Smartmentality and the Social Credit System

Having outlined the development of Hangzhou and Tianjin's SCS, we now turn to three of the mechanisms through which smartmentality works in both cities. These are: the role of calculative digital scoring practices; discourses on moral and behavioural codes and benchmarking of good citizenship; and finally, the geometries of interaction between the state, citizens, and other actors in relation to the SCS.

Calculative digital scoring

Just a few decades ago, very few people could imagine there would be a time when everyone owned their own computer and mobile phone, let alone a smartphone. Computing has now become a constant companion, enabler and lifestyle of urban living. The creation of the internet created a new online virtual sphere intertwined with time and space. This online landscape necessitates, and has developed hand in hand with, ever-larger volumes (and types) of data. Datafication has, at the time of writing, become a key process in online digital life, as well as in specific areas such as the IoT (internet of things).

SCSes rely on datafication, with the datasets that form the core of these systems depending on the aggregation of multiple types of data which are constantly updated. This data is then parsed, analysed, and used based on different algorithms. The ways in which data is collected, evaluated, and interpreted are clearly tied to the politics of datafied urban life. Urban authorities are the key actors who, in the Chinese context, are responsible for interpreting data in specific ways, by generating digital governance logics based on scoring and weighting different activities depending on the particular aims of a given municipal SCS. The algorithmic logics that exist at the heart of SCSes generate credit scores, but they are also drivers for urban residents' engagement with scoring through ordinary urban lives. City residents engage with these logics in both responsive and proactive ways by engaging in activities that may increase (or decrease) their individual credit scores.

In Hangzhou's Qiangjiang SCS, in order to quantify citizens' credit levels, multiple socio-economic data streams are analysed by data analytics-based technologies to calculate scores statistically, thus representing an individual's credit features in the domains of public administration, economic life, judicial contexts, everyday urban life, and non-profit domains. In Hangzhou, credit scores are differentiated into five credit levels (<550: need to improve; 550–600: average; 600–700: good; 700–750: excellent; >750: outstanding). A normal distribution of scores is adopted to ensure there is legitimized differentiation, and reasonable potential to increase scores. Most citizens in the city have credit scores in the 600–750 range. The population with a high score (>750) is seen as representing model citizens with significant social contributions.

In Tianjin eco-city's initial recycling-based SCS, the algorithmic logic was straightforward and clearly defined, and more narrowly focused than

154

in the case of Qianjiang Credit. Scores were determined by the types and amounts of waste recycled by each resident, with this data shared and managed by an urban database and smart city administration centre. In the broader HaiRiver Credit SCS, however, a more sophisticated calculation of credit scores contributes to an algorithmic logic based on: residents' basic information; professional ethics; contractual fulfilment capacity (for example, on-time payment of public utility bills); administrative and judicial credits; social networking and social or community contribution. Thus, a more rounded notion of citizenship is performed through the HaiRiver Credit SCS, representing an evolution from the earlier SCS's exclusive focus on pro-environmental behaviour and the circular economy, to a more complex algorithmic logic through which urban citizenship is constructed and performed through different spheres of everyday life.

Behavioural codes and benchmarking of good citizenship

'With a sincere heart, for xinyi [integrity] things.'
'Credit – the passport of life.'

The above slogans were found on roadside billboards in Tianjin (Figure 8.3). They are expressions of key discourses around SCSes, placing credit scoring at the heart of urban life, and underlining credit (the 'passport' in the slogan above) as the enabler of urban living. They also point to the moral and behavioural incentives and goals of SCSes, such as in the first slogan's prompting for sincerity and integrity.

The above slogans are, more broadly, examples of the ways in which SCSes rely not just on algorithmic logics, but on broader discourses that aim at the production of smart citizens, or in Bridle's (2016) term, 'algorithmic citizenship.' Furthermore, it can be argued that smartness and good credit ratings are not simply technical frameworks and cold calculative processes,

Figure 8.3: The roadside 'advertisements' promoting credit

Source: photo (a) taken by the authors in Hangzhou, (b) and (c) in Tianjin.

but exist within specific social and governance-related structures and subjectivities (Burrell and Fourcade, 2021).

In the SCS context, punishments and rewards are key ways in which smartmentality is operationalized. In Hangzhou, as disclosed by the 2019 Hangzhou Municipal Joint Reward and Punishment Measures Clearance, a list of 302 reward and punishment measures were included in the SCS. These involved 31 administrative departments in Hangzhou. As early as 2016, Hangzhou had issued the country's first joint disciplinary memorandum in the field of transportation, introducing measures for ensuring appropriate conduct in this area. These measures were included in the Hangzhou Municipal Cooperation Memorandum on Implementing Joint Punishment for Untrustworthy Parties with Serious Traffic Violations, and proposed punishments for 29 serious misbehaviours, including driving while under the influence of alcohol and/or drugs. These misbehaviours lead to a decrease in social credit scores.

In the financial arena, Hangzhou's SCS also includes a mechanism for checking individuals' financial trustworthiness before financial decisions are made. This means that the SCS effectively has veto power over residents' applications for finances. Led by the Municipal Development and Reform Commission and the Youth League Municipal Committee, Qianjiang Credit has also been urged to accelerate the implementation of joint incentives for young people to be financially trustworthy and, therefore, creditworthy.

In 2018, Hangzhou city proposed the strategic development of creating a 'deposit-free city', to further the development of the municipal SCS (Qianjiang Credit), and its integration with corporate SCS Sesame Credit. The credit-free programme is structured so that residents with high enough credit scores are deemed trustworthy enough to benefit from 15 deposit-free scenarios and 5 enjoy-before-you-pay scenarios. For example, the public housing rental deposit is halved or exempted for residents with high Qiangjiang Credit scores. Families who are newly allocated to or who are renewing public rented housing contracts can check and validate their Qianjiang Credit at the public rental house service booths. If any member of the family has a score over 700 points, they can apply for a deposit reduction and exemption.

Another part of Hangzhou's SCS is the set of 'Credit-easy+' (Xinyi plus) applications for everyday scenarios. Hangzhou has launched applications such as 'Credit-easy+Parking', 'Credit-easy+Library', 'Credit-easy+Medical', 'Credit-easy+Approval', 'Credit-easy+Loan', 'Credit-easy+Travel', and the like. These provide incentives in specific areas of urban life for those with high scores. The 'Credit-easy+Medical' app, for example, is aimed at helping citizens get medical treatment quickly and efficiently. From July 2019, the city's 245 medical institutes started to provide 'Credit-easy+Medical' services. Patients with high enough scores do not need to pay fees before and during the treatment process, and can pay within 48 hours after the end of the entire treatment process (through self-service machines before leaving the hospital or via smartphones without needing to wait in line). The app also aggregates all medical services that require payment across the city's hospitals into one payment platform, therefore enabling citizens to pay for key urban health services in one go, or at least through the same payment portal. According to municipal data from 2021, since its launch, 1.74 million people have used the app-based service. Patients who benefit from the app spend an average of one hour less waiting in the city's hospitals. Other 'Credit-easy+' services, like smart parking, have connected to the Hangzhou City Brain system, which extends to a broader field of smart urban management and digital governance.

A key aspect of Hangzhou's SCS is the promotion of pro-social behaviour, such as volunteering, through the credits awarded to residents for taking part in these activities. An example of this relates to elderly care. Hangzhou has launched a 'Time Bank' volunteer service as part of Qianjiang Credit. When elderly residents make an appointment for elderly care services, they can check the Qianjiang points of the service personnel they will engage with, and have the opportunity of selecting the personnel who will deal with them based on their credit scores. This effectively performs credit as trustworthiness. However, the system also awards credits to those who volunteer in elderly care-related services: the amount of credit is based on accumulated volunteering time, donations to social causes, and the like.

A further example of the performance of trustworthiness through pro-social behaviour in Hangzhou is the fact that Qianjiang Credit scores can also be increased by participating in activities such as blood donation, other volunteering activities or undertaking making low-carbon travel by bus and/ or subway. In addition to gaining higher scores, these behaviours also result in individuals eventually being awarded specific certificates on the system, such as 'Sports Talent' and 'Low-Carbon Walker' recognitions. These can all be seen as digital platform-enabled examples of state-sanctioned citizenship.

In contrast, the Tianjin eco-city SCS is more clearly defined, with a straightforward focus on recycling and pro-environmental behaviour. The system uses incentives (in the form of credits exchangeable for housewares) as well as disincentives (such as the publication of recycling blacklists on the electronic bulletin board system of communities). For example, in the smart residential community of Keppel-Jijin, the blacklists of recycling and carbon reduction are publicized every month in the 5G Big Data service centre. Thus, even in an SCS such as the eco-city's, where the algorithmic logic is, on the surface, aimed at a single sphere of urban life (namely, pro-environmental behaviour), this single-sector focus is still performed as linked to citizenship. Incentives and disincentives, as well as publically available lists of low-scoring individuals, are ways in which recycling becomes a window and a mirror into a resident's compliance and engagement with citizenship as digitally defined through the SCS platform. This is an example of how smart and platform urbanism can, in specific contexts, become part and parcel of systems of social control (Vanolo, 2014). It is thus key to examine and unearth the discourses and narratives that co-constitute digital social credit.

State-citizen relations in the production of municipal SCSes

Platform urbanism is emerging at the time of writing, but much like other technologies and systems in a technicized society, it is likely that digital social credit will soon become part and parcel of everyday urban life, interwoven with multiple aspects of life, work, consumption, and change in the city. On the one hand, appropriation of data about citizens' behaviour and their digital conduct potentiates a more efficient analysis and management of city communities and urban space. This is undeniable in urban China, where Big Data and algorithmic governance are rapidly developing both in scope and in applications, not just limited to social credit. Indeed, SCSes sit alongside digital health codes (used, for example, for COVID-19 control), and the XueXiQiangGuo platform (an app for Party members for training purposes, including the study of Xi Jinping Thought). Nonetheless, the question remains as to how the state-citizenship relationship is operationalized through SCSes.

It is clearly too early to conclude that Chinese smart urban governance practices, including the SCS, constitute a geographically and nationally uniform system of control without any possibility of pushback. For example, the Qianjiang Credit app is available to every urban resident who works or lives in Hangzhou, as is Tianjin's HaiRiver credit. Yet the recycle-credit program in Tianjin eco-city is only accessible to those who own a house in the area, excluding tenants or other non-permanent eco-city residents. However, overall, municipal SCSes apply a smartmentality based on incentives and penalties for activities that perform (or fail to) state-sanctioned notions of citizenship. The key elements of each SCS include: segmentation of urban life into positive or negative activities; a prioritization of a narrowly defined notion of social harmony over other urban priorities; the use of digital scoring, ranking, and rating practices; making specific scores or individuals publically viewable and held up as 'good' or 'bad' examples; and the sifting of ever more complex data streams about many aspects of individuals' urban life into a single score. The SCS score, in turn, has repercussions on opportunities (jobs, interviews, finance), mobilities and lifestyle (leisure facilities, transport, and the like), and access to services (health, care, public housing). This can be said to recoin what counts as citizenship, and who qualifies as a trustworthy citizen, as well as linking citizenship to active and pervasive data practices. A specific notion of citizenship, then, is segmented through the lens of the SCS into categories, behaviours, and actions. These are then reinterpreted through algorithmic logic into a score that ranks and values each individual as a citizen.

Thus, the above-mentioned operation of SCSes in Tianjin and Hangzhou confirmed the production of specific notions of urban citizenship (Gabrys, 2014; Scannell, 2015), as well as establishing consequences-laden digital envelopes around what qualifies as appropriate conduct. This raises key questions around the continued closure of urban governance, as digital systems such as SCSes introduce yet more complex layers of 'black box' technicity to urban life. However, it cannot be denied that SCSes also represent significant urban and behavioural steering opportunities, which could – in another context perhaps – carry with them the potential for more open, diverse, and inclusive forms of urban life.

Conclusion

In this chapter, the SCS was analysed as an entry point for investigating the evolution of smart and platform urbanism in China, and for engaging with the ways in which urban citizenship is being recast and performed through the datafication of urban lives. In summary, the contribution offered is threefold.

First, municipal SCSes in China can be seen as an evolution of smartmentality. No longer confined to smart cities as geographically bounded projects, smartmentality in the age of platform urbanism can be deployed across cities, 'landing' in each urban location as an iteration of systems (like the SCS) which are the result of national visions for the link between digital platforms and citizens. Analysis of municipal SCSes shows that smartmentality (as mediated through social credit platforms) is articulated through scoring, rankings, and the digital regulation of conduct and behaviour. This, in turn, links urban residents to the governance sphere in ways that turn citizenship into the *performance of 'good' behaviour*. While SCSes differ across cities, this overriding aim is clear across the two cities we considered: citizenship, then, is less based on rights and expectations for each citizen regardless of their 'performance', and more on the duties and deliverable actions that each citizen can be expected to perform, and which urban authorities can (and do) track through smartphones.

Second, our two municipal cases show that the ways in which smartmentality is operationalized in different cities through municipal SCSes respond to national-level steering, while also displaying variations in the emphasis placed on different areas of behaviour, as well as in how 'good' and 'bad' behaviour is defined and rewarded or sanctioned. This helps to underline the point, made at the start of the chapter, that the Chinese SCS landscape is not monolithic, but is a geographically-diverse expression of norms and guidelines for citizenship developed at a national scale for application by more local actors such as municipalities and, to some extent, the private sector. This also underlines another point about the need to consider urban algorithmic governance and the link between smart and platform urbanism and the state in ways that (in the Chinese case, at least) show that state power is as important, if not more, than the power of the market or the drive towards neoliberal logic of urban governance. Thus, considering SCSes in the Chinese case is important not only to understanding urban China, but also to broaden the scope of research on urban data politics more globally.

Third, we have argued that the emergence of the SCS programme is based in part on notions of social and urban harmony, but that this is rooted in a deeper attempt to predict, forestall, and control the development of disorder, whether that be protest or lesser but still 'undesirable' behaviours. Thus, the achievement of an urban and sociopolitical ideal is in large part determined by fears of impending and potential crises. It would be tempting, in this context, to describe the development and complexity of SCSes as hegemonic and unassailable: shining technological monoliths that are so dynamic and intricate that they lie beyond the capacity of any individual or group to resist or influence. The result of this sort of analysis is acceptance of powerlessness, resignation, and lack of agency *outside the bounds of state-prescribed acceptable behaviours*. As a way past this impasse, it is useful here to draw on de Certeau's notion of *tactics* through which the everyday can be a site for challenge and resistance (Frow, 1991; Silverstone, 1989). In his work, de Certeau underlines how tactical moves in time can be seen as a response to the *strategies* put in place by the state to exert dominion – spatially and in other ways (de Certeau, 2005). The SCS is a key example of a strategy which is developing – initially in a piecemeal way, city by city, province by province, but whose overall aim tends towards strategic integration and enhanced and spatialized control. Through tactical and long-term practices such as reworking of (outwardly) dominant and hegemonic systems such as SCSes, or through the quiet and subversive utilization of SCSes in ways they were not intended to be used in, there is the potential for SCSes to be reinterpreted and challenged in ways that do not equate with frontal resistance of the type that would incentivize crackdown. What might these tactics look like? That is the next step in researching on the plaformization of digital urban society in China, an endeavour requiring deep ethnographic work in engaging with the everyday.

Acknowledgements
This work was supported by the Chiang Ching-kuo Foundation (RG009-U-18).

References

Bridle, J. (2016) 'Algorithmic citizenship, digital statelessness', *GeoHumanities*, 2(2): 377–81. https://doi.org/10.1080/2373566X.2016.1237858

Burchell, G., Gordon, C. and Miller, P. (eds) (1991) *The Foucault Effect: Studies in Governmentality*, Chicago: The University of Chicago Press.

Burrell, J. and Fourcade, M. (2021) 'The society of algorithms', *Annual Review of Sociology*, 47: 213–37. https://doi.org/10.1146/annurev-soc-090 820-020800

Caprotti, F. (2014) 'Critical research on eco-cities? A walk through the Sino-Singapore Tianjin Eco-City, China', *Cities*, 36: 10–17. https://doi. org/10.1016/j.cities.2013.08.005

Caprotti, F. (2019a) 'Authoritarianism and the transparent smart city', in C. Lindner and M. Meissner (eds) *The Routledge Companion to Urban Imaginaries*, London: Routledge, pp 137–46.

Caprotti, F. (2019b) 'From Shannon to Shenzhen and back: Sustainable urbanism and inter-city partnerships in China and Europe', in X. Zhang (ed) *Remaking Sustainable Urbanism: Space, Scale and Governance in the New Urban Era*, Singapore: Springer, pp 101–19.

Caprotti, F. (2019c) 'Spaces of visibility in the smart city: Flagship urban spaces and the smart urban imaginary', *Urban Studies*, 56(12): 2465–79.

Caprotti, F. and Liu, D. (2020a) 'Emerging platform urbanism in China: Reconfigurations of data, citizenship and materialities', *Technological Forecasting and Social Change*, 151. https://doi.org/10.1016/j.techfore.2019. 06.016

Caprotti, F. and Liu, D. (2020b) 'Platform urbanism and the Chinese smart city: the co-production and territorialisation of Hangzhou City Brain', *GeoJournal*, 87: 1559–73. https://doi.org/10.1007/s10708-020-10320-2

Caprotti, F., Chang, I.-C.C. and Joss, S. (2022) 'Beyond the smart city: A typology of platform urbanism', *Urban Transformations*, 4. https://doi.org/ 10.1186/s42854-022-00033-9

Cardullo, P. (2020) *Citizens in the 'Smart City': Participation, Co-production, Governance*, London: Routledge. https://doi.org/10.4324/9780429438806

Cardullo, P. and Kitchin, R. (2019) 'Smart urbanism and smart citizenship: The neoliberal logic of "citizen-focused" smart cities in Europe', *Environment and Planning C: Politics and Space*, 37(5): 813–30.

Crampton, J.W. and Elden, S. (eds) (2007) *Space, Knowledge and Power: Foucault and Geography*, Aldershot: Ashgate.

Credit China (2018) 'The demonstration cities of credit construction: Hangzhou, Zhejiang', Credit China, [online] 11 February, available from: https://www. creditchina.gov.cn/chengxinwenhua/chengshichengxinwenhua/201802/ t20180211_108736.html

Creemers, R. (2018) 'China's Social Credit System: An evolving practice of control', [online], available from SSRN: http://dx.doi.org/10.2139/ssrn.3175792

Cugurullo, F. (2021) *Frankenstein Urbanism: Eco, Smart and Autonomous Cities, Artificial Intelligence and the End of the City*, London: Routledge.

Dai, X. (2020) 'Toward a reputation state: A comprehensive view of China's Social Credit System project', in O. Everling (ed) *Social Credit Rating: Reputation und Vertrauen beurteilen*, Wiesbaden: Springer Gabler, pp 139–63. https://doi.org/10.1007/978-3-658-29653-7_7

de Certeau, M. (2005) 'The practice of everyday life: "Making do": Uses and tactics', in G.M. Speigel (ed) *Practicing History: New Directions in Historical Writing after the Linguistic Turn*, London: Routledge, pp 217–27.

Duffie, D., and Economy, E. (eds) (2022) *Digital Currencies: the US, China, and the World at a Crossroads*, Stanford, CA: Hoover Institution Press.

Ellul, J. (1962) 'The technological order', *Technology and Culture*, 3(4): 394–421.

Fields, D., Bissell, D. and Macrorie, R. (2020) 'Platform methods: Studying platform urbanism outside the black box', *Urban Geography*, 41(3): 462–68.

Frow, J. (1991) 'Michel de Certeau and the practice of representation', *Cultural Studies*, 5(1): 52–60.

Gabrys, J. (2014) 'Programming environments: Environmentality and citizen sensing in the smart city', *Environment and Planning D: Society and Space*, 32(1): 30–48. https://doi.org/10.1068/d16812

Grossi, G. and Pianezzi, D. (2017) 'Smart cities: Utopia or neoliberal ideology?', *Cities*, 69: 79–85. https://doi.org/10.1016/j.cities.2017.07.012

Jinwanbao (2021) Small number, big data and zero-waste city. *Tonight News Paper (Jinwanbao)*. Tianjin: China-Singapore Tianjin Eco-city Administrative Committee, available from: www.eco-city.gov.cn/p1/stcxw/20210517/43460.html

Joss, S., d'Assenza-David, H. and Serra, L. (2022) 'Eco-neighborhoods and the question of locational advantage: A socio-spatial analysis of French "ÉcoQuartiers"', *Cities*, 126. https://doi.org/10.1016/j.cities.2022.103643

Kitchin, R., Cardullo, P. and Di Feliciantonio, C. (2019) 'Citizenship, justice and the right to the smart city', in P. Cardullo, C. Di Feliciantonio and R. Kitchin (eds) *The Right to the Smart City*, Bingley: Emerald, pp 1–24.

Kitchin, R., Coletta, C. and McArdle, G. (2017) 'Urban informatics, governmentality and the logics of urban control', Programmable City working paper 25, [online], available from: https://osf.io/preprints/socarxiv/27hz8/

Müller, M. and Trubina, E. (2020) 'The Global Easts in global urbanism: Views from beyond North and South', *Eurasian Geography and Economics*, 61(6): 627–35.

Pow, C.P. (2017) 'Sensing visceral urban politics and metabolic exclusion in a Chinese neighbourhood', *Transactions of the Institute of British Geographers*, 42: 260–73.

Rodrigues, N., Vale, M. and Costa, P. (2022) 'Urban experimentation and smart cities: A Foucauldian and autonomist approach', *Territory, Politics, Governance*, 10(4): 549–67. https://doi.org/10.1080/21622671.2020.1777896

Rose, N. (1999) *Powers of Freedom: Reframing Political Thought*, Cambridge: Cambridge University Press.

Scannell, J. (2015) 'What can an algorithm do?' DIS Magazine, [online], available from: http://dismagazine.com/discussion/72975/josh-scannell-what-can-an-algorithm-do/

Shen, M. (2020) *Rural Revitalization through State-Led Programs: Planning, Governance and Challenge*, Singapore: Springer.

Silverstone, R. (1989) 'Let us then return to the murmuring of everyday practices: A note on Michel de Certeau, television and everyday life', *Theory, Culture & Society*, 6(1): 77–94.

Törnberg, P. (2023) 'How platforms govern: Social regulation in digital capitalism', *Big Data & Society*, 10(1). https://doi.org/10.1177/205395 17231153808

Vanolo, A. (2014) 'Smartmentality: The smart city as disciplinary strategy', *Urban Studies*, 51(5): 883–98. https://doi.org/doi:10.1177/004209801 3494427

Wong, T. (2022) 'Henan: China Covid app restricts residents after banking protests', BBC News, [online] 14 June, available from: https://www.bbc.co.uk/news/world-asia-china-61793149

Xu, Y., Caprotti, F., Zhang, W. and Pan, M. (2022) 'The socioenvironmental state and urban transitions: Eco-urbanism in China and the UK', *Environment and Planning E: Nature and Space*. https://doi.org/10.1177/2514848622 1132835

Zhang, C. (2020) 'Governing (through) trustworthiness: Technologies of power and subjectification in China's social credit system', *Critical Asian Studies*, 52(4): 565–88.

Zhang, J., Bates, J. and Abbott, P. (2022) 'State-steered smartmentality in Chinese smart urbanism', *Urban Studies*, 59(14): 2933–50. https://doi.org/10.1177/00420980211062888

Zhang, P., Zhang, K. and Yang, N. (2022) 'Fears of data abuse as Chinese health code turns red for financial scandal protesters', *South China Morning Post*, [online] 14 June, available from: https://www.scmp.com/news/china/science/article/3181635/chinese-health-code-turns-red-financial-victims-about-protest

Zhao, W. and Zou, Y. (2021) 'Smart urban governance in epidemic control: Practices and implications of Hangzhou', *Chinese Public Administration Review*, 12(1): 51–60. https://doi.org/10.1177/153967542101200104

PART III

Tactics

9

Platform Work, Everyday Life, and Survival in Times of Crisis: Views and Experiences from Nairobi

Prince K. Guma

Introduction

The COVID-19 pandemic highlighted the impact and transformation of data and data-driven tools in the supply, provision, and delivery of goods and services, and how entangled these systems are in urban lives, livelihoods, and the modes by which cities are rendered liveable by their inhabitants in times of crisis. In Southern cities, digital platforms have been considerably at the heart of urbanization processes (see, for example, Guma, 2022b), filling certain voids during moments of lockdown. They have been integral for bypassing infrastructural obduracies of the pandemic crisis, and reforming and reconstituting everyday and mundane processes in the city such as marketing, transmitting, supplying, and consuming goods and service. Digital platforms have also facilitated new models and innovations for making orders and payments for goods, services, and products. The ubiquity of digital platforms and innovations including billing and payment for goods and services, financing and crediting, consumption and tracking, and querying facilitate interaction between service providers and consumers. They have come to define the new realities and reconfigurations of everyday urban processes, provoking new ways of thinking about emerging constellations of actors, networks, infrastructures, resources, and technologies being formed (Lee et al, 2020). The new critical engagements being enabled, systems of power being organized and exercised, new formations of urbanity being mobilized, and infrastructure renaissance or remaking being actualized all constitute a way of future in the post-pandemic city.

In this chapter, I examine articulations of platform work, everyday life and survival in times of crisis. I offer an expanded approach to understanding articulations of data and data-driven solutions in terms of their everyday and mundane manifestations during the COVID-19 pandemic. While there has been recent focus on the aspect of platform urbanism and data-driven solutions, there remains a dearth of research examining digital platforms beyond the usual tripartite demarcation of the social, economic, and political, and not least registers of technology (that is, flows and distribution of technology), finance (fintech and flows of financial capital), media (media channels and startups), and governance (mediums for governing) (see, for example, Edward, 2020; Lee et al, 2020; Scheepers and Bogie, 2020; Sitas et al, 2022). Most accounts are situated within dominant corporate discourses and examine data-driven platforms as business-led initiatives that channel urban development strategies toward quasi-unavoidable ICT-driven urban futures (ibid.). Often, studies tend to take a determinist view that does not grant urban stakeholders outside the corporate framework much agency in shaping digital presents and futures. Yet cities do not necessarily always adopt the circulating models of urbanization and as can be seen in many Southern cities where digital platforms have become central to everyday life (see, for example, Guma and Mwaura, 2021), they are increasingly being shaped by user-orientated knowledges, appropriations, and practices that are deeply ingrained in situated world-views and activities that play an integral part in how residents inhabit contexts of urban heterogeneity.

Thus, in light of platform work, everyday life, and survival amidst crisis, it is important to examine the types of engagements and operations evident in the use and appropriation of digital platforms during the COVID-19 pandemic. In answering this question, I make an explicit contribution that is empirically grounded in the need to better understand situated realities and urban politics and rationalities of delivery platforms in a Southern city. I draw from my own broader research work and encounters with a wide range of digital infrastructures in Nairobi (Kenya's capital, a political and economic hub, and a major commercial, diplomatic, technological, and cultural centre) between March 2020 and March 2022, and critical reflections and extensive deliberations relating to the ways through which platform work, everyday life, and survival sizzles and congeals at the intersection of the pandemic restrictions, and the urban and infrastructural responses to the COVID pandemic in Nairobi.

Building on established debates on urban and infrastructure development and appropriation (see, for such debates, Guma, 2022c; Guma and Monstadt, 2021), this chapter makes a contribution that is empirically grounded beyond utopian descriptions of circulating techno-centred visions and deterministic views of urban innovation. I foreground lived and actual arrangements of digital infrastructure through a combination of everyday coping strategies and mechanisms of hand-to-mouth survivalism, smart improvisation, frugal innovation, and creative calculation of risk at the height of COVID-related socio-spatial inequalities. By centering basic practices that exist beyond formalities of the formal markets and corporate culture, I bring into discourse authentic realities and contexts of the South of the technologies and their bifurcation with the urban informal sector and economy.

Conceptually, I derive inspiration from postcolonial critique of urban and technology studies to offer a more nuanced and creative way of thinking about technology, infrastructure, the city, and society, one that considers technological transformations as always in flux, open to tappings and trappings, and 'always relative to the different constituencies, populations, and agencies at work' (Dourish and Bell, 2011, 194). I reiterate Sheller's point that technology does not enter people's lives as a 'black box' and 'neutral set' (2004, 208) but, as always, needs to be opened up to exploit the 'array of social actors, processes and images therein' (Steen, 2011, 1). Therefore, I provoke a more dynamic, open, and relational view of digital platforms, especially one that imagines digital platforms not just as manifestations of urban entrepreneurialism and a technocratic development agenda but one that acknowledges its context-specific appropriation. The point of departure from which I set off is the premise that digital technologies cannot be viewed as a decontextualized incident. Neither can they be understood as entirely top-down phenomena. Instead, they are part and parcel of institutional patterns that manifest through situated processes and practices of translation and appropriation – thus, shaped by social and contextual realities. Therefore, it is particularly important to go beyond assumptions that citizens, cities, and technologies, despite their diversity, possibly will intersect through sensor-activated programs (Guma, 2022b, 2022c) to realize seamless sets of material and non-material relations and perfectly orderly, complete, and immanent systems as the driving force of urbanization in the digital age. I contend that cities are rather characterized by situated contingencies, nonlinear progressions, and transient temporalities that evolve across human and more-than-human registers. Therefore, this chapter offers a critical view beyond dominant corporate discourses on business-led initiatives that channel urban development strategies, toward an ethnographic view that centres on human capital of platform urbanism flows that highlights alternative platform-driven urban futures.

The remainder of this article is structured as follows. First, I highlight different ways through which operational companies had to make room for digital systems during the COVID-19 restrictions. Second, by way of 'following the platform' from a worker and a driver operating in fundamentally unstable, indelibly precarious urban contexts of Nairobi, I offer ethnographic stories of Mariam and Alex to describe how at the height of COVID-related socio-spatial inequalities, residents appropriated digital systems for survival through their embodied and empirically grounded everyday practices and contexts. Third, I argue that the interactions between diverse platforms and their various users, including platform workers, create what Simone (2004) has referred to as 'people as infrastructure'. I highlight what different tactics and mechanisms are employed as mediums to navigate life in times of crisis. In concluding, I offer reflections about what the entanglement of bodies, infrastructures, and platforms through everyday life and survival could mean for planning and theorizing the 'post-COVID' city and city of the future.

Making room for digital systems during COVID restrictions

In many Southern cities, the COVID pandemic exposed, if not exacerbated systemic and institutionalized socio-spatial inequalities. In eastern Africa, the Kenyan government deployed far-reaching measures and directives to control transmission and mitigate the social and economic impacts of the pandemic. It put in place a number of public health measures, including travel bans and closures of borders, schools, workplaces, open markets, entertainment spots, and places of worship. It also implemented evening curfews and mandatory quarantines, authorized increases in health service capacity and supplies, and expanded mass testing in several cities in the country (Guma, 2022a). A lot of these measures and restrictions were incremental, with evening curfews being the most consistent. The government endeavoured to sensitize the public through different forms of media as it set up portals, opened toll-free lines and WhatsApp channels to disseminate reliable information on the pandemic as well as enable citizens to report suspected cases (Guma 2022a). The state's 'technological solutionism' could also be witnessed in its deployment of similar applications and platforms to monitor movement of citizens, restrict access to public spaces, and impose work-from-home rules. In the process, such developments further exacerbated the struggles of many residents who were already affected by the exclusive nature of urban infrastructures and technologies.

As the lockdowns and curfews led to several restrictions related to the movement of goods and services, operational companies had to close their stores and businesses to the public and deploy digital platforms in their daily operations as a way of navigating the restrictions. Partially because of COVID-related challenges and restrictions, urban sectors sought digital technological solutions in the urge to improve service delivery instruments and capacities, and expand their own market share and revenue streams in times of crisis. Survival or 'staying afloat' for many such companies meant quickly adapting to compete fiercely in clouded and ambiguous markets by 'making room' for mobile-phone based options, like calling and text messages, or launching a new app. Many sectors deployed these technologies as a magic bullet to the crises caused by the COVID-19 pandemic. Digital platforms became leveraged by a wide range of actors for reorganizing urban spaces and service delivery, with these processes signalling unprecedented shifts in which they deployed exciting innovations whose very essence and success were determined by the use and apparent influx and saturation of the mobile phone in cities. New models, services, products, and innovations were evident and they shaped dynamical transformation as well as a kind of new value creation and disruption in many sectors. With regards to service provisioning, for instance, different kinds of mobile technological innovations were radically mobilized for supply, provision, and delivery. While some companies developed their own platforms, others used third-party or shared platforms where they did not necessarily own rights to the infrastructure of the platform(s). Most companies signed up to dominant digital platforms that combine e-commerce and ride-hailing to offer third-party services. In fact, it was simply implausible for companies to operate in the market without deploying digital platforms during the pandemic. With some companies increasingly deploying communication services that include calling, SMS texting, and use of web browsers, others also deployed mobile phone-based applications and systems for payment relying on text and short code through encrypted SMS (short messaging service) and USSD (unstructured supplementary service data) platforms accessible with limited or no broadband capabilities. This made e-commerce become particularly convenient and prominent, providing critical continuity for service provision and delivery in the midst of COVID-19 related restrictions. In other words, the pandemic showed the implications and consequences of state agency and politics to companies in the digital age.

In tandem with these developments, technology companies launched different offers that made it easy for businesses and start-ups to sign up. For example, Uber waived activation fees for new restaurants, which made it quicker and easier for new restaurants to join UberEats; reduced wait times to less than 24 hours for new sign-ups; launched a new feature, giving restaurants the option to receive daily (rather than weekly) pay-outs, in order to help with cash flow reliability during those uncertain times; and partnered with various stores to offer customers essential items such as over-the-counter medicines, toiletries, and essential household items (Scheepers and Bogie, 2020). For example, food companies and restaurants or delivery stores that resorted to takeout orders registered with and joined UberEats, where clients could simply order their meals and drinks virtually, hiring platform workers and offering bonuses and incentives for deliveries. Consequently, demand for boda boda (a two-wheeled motorcycle-taxi powered by an internal combustion engine and constituting an affordable mobility option designed 'for hire' of passenger taxi services and movement of light goods and non-palletized items such as cargo) as a mode of transport and delivery soared during COVID restrictions. Boda bodas quickly became common means of public transportation and delivery services, establishing themselves in the transport sector as a micro-mobility service providing the largest, most dynamic and convenient means for moving and transporting people, goods, and services amidst increased vulnerability to varied forms of police violence, new threats, security risks, and violence (Figure 9.1).

Motorcycle taxis took on the roles of delivering goods and services, shaping demand-responsive transit services in the process. With these taxis providing a means to connect the demand and supply sides, demand for them increased as residents opted to receive meals, groceries, and other products like parcels straight to their households, with motorcycle taxis becoming pre-eminent modes for providing door-to-door services, with these constituting a new business model and a new work order for the taxis (see, for example, Sitas et al, 2022; Guma, 2022a, 2022b). During pandemic restrictions, urban residents embraced digital mobility and delivery platforms to enable virtual orders and home delivery services, further catalysing the growth of new platforms (Figure 9.2). Subsequent sections interrogate everyday use of digital platforms during COVID-19, with the aim to introduce the reader to urban and situated forces and realities of platform work and challenges in times of crisis. Following the platform from Mariam as a worker and Alex as a driver, I demonstrate how digital systems and delivery platforms have been employed to navigate urban life, survival, and challenges of pandemic restrictions.

Figure 9.1: A boda boda stage in an open market by the roadside

Source: image taken in Nairobi by author, in October 2021.

Figure 9.2: An outdoor advertising billboard in a high-traffic area

Source: image taken in Nairobi by author, in October 2021.

Platform work, everyday life, and survival: through the lens of Mariam and Alex

The difficult conditions of COVID restrictions and lockdowns, which meant that different companies such as hotels, fast food restaurants, and cafes laid off workers, shifted to working remotely to 'stay afloat', or signed up to existing digital platforms (both SMS- and text-based platforms that require limited to no broadband capabilities, and the more advanced and complex platform labour companies like Uber and Bolt), had implications for employees who lost their jobs or had to 'work from home'. Mariam, who worked at a high-end hotel before COVID, is one of the many employees who was laid off at the height of pandemic restrictions. She had to face the exceedingly difficult conditions of uncertainty – at home under desperate economic conditions, sometimes without the ability to afford vital services like food, water, sanitation, and energy. Tired of waiting for things to get 'back to normal', Mariam was left with no option but to seek alternative modalities of care and interdependence as a form of critical social infrastructure. In her urge for survival, Mariam started a takeout business, signing up with digital platforms that combine e-commerce and ride-hailing to deliver her food products. Mariam opted to compete in what was then an emergent but highly ambiguous market of food delivery. This business became particularly convenient for Mariam. It provided critical continuity in her line of work (the kind of work she had lost at the hotel).

173

While using social networking sites like Instagram, Twitter, WhatsApp, TikTok, Snapchat, Facebook, and YouTube, Mariam sought to create content and keep up and promote her business, publishing articles, videos, photos, and all kinds of items in the range of online streaming, social media, and entrepreneurship to survive in the business and sustain her livelihood as well as her own competitive advantage against the bigger and more established stores and restaurants in the market. In part, she would do so strategically to create, sustain, and diversify her own revenue sources through incremental sales, but also by improvising, innovating, and creatively calculating risk. Mariam also employed available digital platforms (Uber and Bolt) and mobile payment systems (M-Pesa and Airtel Money) in her daily work. These digital technologies offer low-cost marketing opportunities within a city where survival demands such kinds of tactics and calculations where one must manoeuvre to make ends meet. They allowed Mariam a cost-effective means to not only interact with potential customers and clients, but also to create new markets, navigate profit-making gaps, and sustain herself in business.

Beyond Mariam's use of such platforms, she also relied on friends, family, and colleagues, including Alex, a boda boda driver. In Mariam's most desperate times, Alex came in handy to make the deliveries on her behalf, often on short notice. Alex lives in Kawangware, a low-income residential area in Nairobi about 15 kilometres west of the city centre, close to Lavington, a diverse and high-income residential estate with different restaurants and hotels, including the one where Mariam worked before the pandemic. In Kawangware, Alex lived with his family in a single partitioned flat. For Alex, this arrangement with Mariam was ideal as it exists within the realm of a kind of social infrastructure that is highly symbolic of a continuation of the traditional and long-established practices founded on notions of mutual assistance that have allowed residents in communities like Kawangware to manage realities of scarcity of urban services. Largely based on the worldviews of sharing and community, such mechanisms embody relational, collectivist, and intuitive ideals of community.

Before COVID, Alex worked on a 'stage' located close to where Mariam resides. However, Alex could not work at this stage any more due to COVID restrictions and lockdowns, as these meant that Alex had to adjust his modes of operation. For instance, he would be expected to own and use a smartphone for on-demand deliveries (Figure 9.3). He would be expected to connect to reliable internet and use pre-paid data packages. He would be expected to sign up with dominant platform labour companies and ride-hailing and delivery apps like Bolt or Uber mostly to make deliveries. Even worse is that he would be expected to meet the requirements of these dominant companies. Digital mobility platforms mandate requirements that are unrealistic to drivers, as they tend to assume a degree of stability for the drivers, thus barely capturing the social reality of the boda boda industry.

Figure 9.3: A boda boda operator processing and transmitting a delivery at a store by the street

Source: image taken in Nairobi by author, in October 2021.

For example, signing up to dominant ride-hailing and delivery apps like Bolt or Uber Boda requires one to provide documents that many motorcycle drivers often do not have – these include providing evidence of a motorbike driving licence and national ID card; providing evidence that the motorcycle is in new and therefore good condition; providing evidence of a passenger service vehicle insurance and motorbike logbook, a sales agreement photo for the motorbike, and a police clearance certificate. Moreover, in addition to all of these, one is also required to obtain two reflectors and two helmets. The other requirements expected of the drivers, such as the necessity for smartphone ownership, use of data packages, and need for reliable internet connection, means that motorcycle-taxi drivers cannot be illiterate.

While Alex found (mobility) platforms exciting as they offered a new work opportunity for his occupation and allowed him to work with flexible schedules, stops, and pricing, their requirements assumed a degree of stability that does not capture the social reality of people like him in the industry in which he operates. For example, mobility operations took place under a lot of pre-existing institutional voids in regulations, and were largely unregulated, or governed by laws that barely anticipated the model through which digital services were used. This means that the new laws and regulations for the most part were barely embraced by boda boda operators and drivers. Boda boda drivers like Alex didn't even try to adhere to them, leading to very low compliance. As Alex argued, boda boda drivers use all means possible to evade the law, leading to generalized lack of adherence to safety prescriptions that fosters dodgy practices: "Once an operator acquires motorcycles, they want to start the business; so they revert to the daily routine of scrambling for customers while flouting traffic rules without diligence of the laws or order."

Indeed, Alex operated at the intersection of multiple marginalities – spatial, infrastructural, legal, regulatory, and socio-economic: constantly aware that if he was to make ends meet, he had to take the required risks where necessary. Such risks fell along the spectrum of formality and informality, legality and illegality. For instance, he would 'cut corners' when he had to; take the back route when he had to; freely seek counter theses to state and platform regulations that he deemed prohibitive. That is how his work life was configured: to evade surveillance, to bypass dangers of reductive policies, and to counter risk. This is precisely because cutting corners through the 'back route' is the natural orientation of many residents in Nairobi, who are always seeking room for manoeuvre. This mentality leads to insurgent and counter-institutional forms of operation, and forms the material condition of his and other boda bodas' operations in the city, which ultimately becomes implicated in the production of pandemic urban space. Unlike vehicular modes of transport that disregard the daily engagements, negotiations, and relations that comprise makeshift urbanism, boda bodas circumvent and bypass official routes as well as infrastructural vulnerabilities, inadequacies, and absences, to navigate back routes locally christened as 'panya routes' (also known as backroad infrastructures that diverge from the main route kinds).

Within this context, drivers like Alex had to deal with the complex dynamics in highly precarious contexts, where survival became an urban way of life. As Alex argued, he was often aware that he, like most drivers in the city, had to sell their labour to digital platforms often with neither legislation from the state, nor social protection, job security, and fair pay. While their legal and employment status and safety standards vary greatly between operators, platform companies leave workers largely unprotected by employment and labour laws, shifting more of the burden of economic risk onto them. Thus, platform workers, as freelancers, face the challenges of managing their own health (during the pandemic) and financial risks (due to the nature of their financial situation), and so on. Besides, the algorithmic controls and tracking systems through which drivers are disciplined, monitored, and surveilled are all hidden from the drivers and platform workers. This becomes an issue for some drivers who remain precarious workers, selling their labour to digital platforms in a landscape where legislation, social protection, job security, and fair pay are either nonexistent or at the bare minimum. As a freelancer, Alex faced the challenges of managing his own health and financial risks during pandemic restrictions. Such questions were important for Alex, especially in the post-pandemic era where boda boda operations that provide the infrastructure for digital mobility and delivery were, in fact, no longer a peripheral means of transport but right at the centre of urban mobility and transport.

What seemed obvious during the COVID pandemic was how motorcycle taxis provided complementary 'gap filling' services to the more traditional modes of transporting and delivering goods and services in Kenya, forming a kind of backbone of the transport passenger and delivery system in the city. As gap fillers, boda boda taxis provided a large source of employment for the ever-growing youth population – where economically driven Kenyans, in their quest for survival labour, drive through complex paths and steep terrains of cities, townships, and villages, carrying luggage and conveying people to their destinations, and repair and maintain the vehicles as assemblers, mechanics, welders, and cleaners. The intrinsic characteristics of platform workers like Alex suited well the realities and constraints imposed during the pandemic, such as lockdowns, when they emerged to fill the gap of delivery from door to door. Food providers, increasingly faced with ordering experiences requiring home deliveries, led to the increased popularity and prominence of boda bodas facilitating phone orders and restaurant deliveries during the lockdowns and pandemic restrictions.

The experience of Alex's activities, practices and modes of operation highlight the equivocal role of the state in maintaining informality. It was clear during the COVID pandemic that the regulatory environment remained highly incomplete, especially for digital mobility and delivery platforms, which rose at the margins of the law during the global pandemic. However, this does not discount plenty of policies and regulations in place that guided the operations of platforms and their companies, motorcycle drivers, and the end users or consumers. Yet public authorities in practice relinquished the implementation of rules to the operators and drivers themselves, as well as to associations or unions. The aim for public authorities, it seems, was less to improve the operating conditions of motorcycle taxis than to limit their role or eradicate them. Take the fact that boda boda taxis are not legally recognized as public service vehicles (Olvera et al, 2020), and Kenya lacks a policy framework in place to register motorcycle taxis as public service vehicles equivalent to what the Transport Licensing Board does with the registration of 'matatus' (essentially small-scale privately owned minibuses in Kenya licensed to carry 14 passengers): a common form of transportation and essentially a shared taxi.

Alex's experiences and those of Mariam highlight creative practices of making do and getting by, and they provoke us to think about the post-pandemic city through the plight and potential of populations at the interstices of planning and city-making. They highlight how urban and infrastructural reconfigurations are incrementally produced through material configurations (Silver, 2014), as hybridized forms of socio-technical production (Furlong, 2014), and as ad hoc, improvised practices that create vast possibilities (Simone, 2004). Alex and Mariam's experiences speak to the agency of critical consciousness, opportunistic knowledge, and frugal technologies and infrastructures at the height of a pandemic. Their stories highlight different undertakings, operations, and navigations of urban life and platform work that demand survivalist tactics.

Platform work as infrastructure in heterogeneous urban contexts

The interactions between diverse platforms and their various users, including platform workers, create an infrastructure that Simone's essay referred to as 'people as infrastructure' (Simone, 2004). To avoid romanticizing these intersections, Simone (2016) reminds us of the intensely politicized mixing forces that constrain and stretch survival, an angle which Doherty (2017) has highlighted to capture infrastructural violence and the aspect of 'disposable people as infrastructure'. These flexible human infrastructures enable the drivers to carve new routes and networks on these routes by figuring arrangements of configured urban materials and networks available to their advantage (Simone, 2008; 2016; Doherty, 2017). These are important for opening up space to alternative conceptions of urbanity, especially those that foreground lived experiences and implications within situated urban localities, particularly in a way that actually highlights the shifting materialities and new circulatory assemblages and representational tropes of urban life – which have also been theorized as incremental (Silver, 2014), lively (Amin, 2014), and incomplete (Guma, 2022c). These are important for analysing digital platforms as tools, networks, and systems that are heterogeneously entangled.

Within the Global South, such analysis has broader implications for platform work as infrastructure in three main ways. First, urban residents are always bound to rely on alternative, off-grid, and makeshift solutions by private actors where such solutions are destined to constitute back-up, distant, leapfrog solutions that – while seemingly unstable, unlicensed, and private – may constitute a mode of operation in such heterogeneous settings. In the same fashion, it ought to be remembered that Nairobi is a highly diverse and splintered city where informal settlements are ubiquitous. It is a city that continues to grow beyond its more than 4.55 million inhabitants (Kenya National Bureau of Statistics, 2019) in tandem with rural-to-urban migration and a rise in informal settlement. For example, it is estimated that 60–70% of the city's residents reside in low-income urban areas that occupy only 5% of the city's residential land (see Olima, 2001). Within such a city, persistent barriers of access to reliable services become definitive and accentuate considerable challenges of inequitable, exclusive, and ill-fitted service supply. Emergent operations moderated by data and data-driven tools and platforms begin to constitute a rather more generalized mode of urbanization as they become crucial for navigating sporadic urban and infrastructural difficulties. The infrastructural configurations of such tools and platforms tend to be fluid, rather than being fixed, constituting creative manoeuvres shaped by processes and practices in which residents negotiate different ways of dealing with urban systems. The example of Mariam in particular highlights the increasing web and flux of creativity, ingenuity, inventiveness, innovation, and experimentation that attract and facilitate a surge in the growth of the informal or 'kadogo' and 'jua kali' (Kenyan slangs for informal, provisional, and small or small-scale) economies in Nairobi.

Second, urban livelihood and survival becomes a thickening of heterogeneous strategies into webs of networks of social infrastructure where digital systems play an integral role in allowing the urban poor to exercise their tactics in different ways to make ends meet. The story of Mariam highlights how during times of crisis, residents increasingly seek out opportunities and creative responses comprising a range of humanistic practice, such as relying on each other for support, and interreliance to counter immeasurable and almost constant infrastructure vulnerabilities. These articulations highlight people as infrastructure, and how through innovative and creative solutions they devise solutions in real time among themselves in part, to counter overdetermined infrastructural and governance stacks and challenges. In particular, they highlight innovative ways of appropriating mobile telephony in Africa (Cardon, 2005; Chéneau-Loquay, 2010). What becomes even more visible through such articulations – particularly for the digital systems being appropriated by residents like Alex and Mariam – is that platform logics offer new modes of operation and new sets of rules and strategies that make possible different kinds of détournement, improvisations, and creativity. Their materialization makes such humanistic and social webs and networks possible, where residents are able to then rely more on social infrastructure or 'people as infrastructure', in a way allowing residents to perform their agency and express their intentions without necessarily being constrained by the extractive, capitalistic, and often hegemonic nature of digital platforms.

And third, urban residents are always bound to interpret hegemonic and often constrictive standards and excessive regulations (including pandemic restrictions and lockdowns) as censorship and to find routes around them through variegated ways that transcend legal and illegal, formal and informal, and order and disorder binaries. Residents are bound to employ crafty and exceptional actions and behaviours to navigate challenges. Across Nairobi, where the majority population owns a smartphone and where more than half the population has access to internet connectivity. For even some of the most vulnerable groups, a mobile phone becomes an essential tool for navigating standards and formalities in unimaginable ways in order, for instance, to access services that would otherwise have been inaccessible or even unaffordable. Frugal technologies like mobile phones and apps helped urban residents who had been further pushed into precarity during the pandemic restrictions to connect and gain access to critical services and needs. These technologies, in addition to disaggregated infrastructures like motorbikes, helped residents to navigate and circumvent the regulations implemented by the state, liberating them from both place and group, and by extension, granting freedom from worlds or rather realities of fixity and obduracy. The stories of Alex and Mariam highlight the ambivalence of urban survival and ways through which residents' webs of survival activities provisionally intersect with technology and infrastructure, and where technologies and infrastructures evolve in flexible and radically open ways depending on residents' ability to engage with their complexity.

Conclusions and considerations for research

This chapter sought appropriations of data and data-driven platforms by way of 'following the platform' from a worker and a driver operating under the auspices of heterogeneous urban contexts. By offering a postcolonial perspective on precarious work during and through the pandemic, and ethnographic stories of how at the height of COVID-related socio-spatial inequalities residents appropriate different digital platforms to navigate urban problems and restrictions, I sought to highlight the relational nature of digital platforms mediated by urban residents in contexts of urban heterogeneity to make an explicit contribution that is empirically grounded beyond utopian descriptions of circulating techno-centred visions and deterministic views of urban innovation. I have demonstrated how digital platforms intersect with bodies of different population groups inhabiting urban worlds and environments in the Global South. With increasing urbanization, digital platforms and innovations are likely to increase in the near future. They open up new questions regarding what lessons we should take from the multiple relations and intersections between technologies, infrastructures, bodies, and the everyday to scale up policy and planning for sustainable post-COVID urban futures.

Considering the diverse and splintered context of the urban South, ordinary accounts such as Alex's and Mariam's are important to consider: first for the flexibility of human infrastructures in carving new routes and networks, and configuring urban materials and networks available to their advantage (Simone, 2008; 2016; Doherty, 2017); and second, in planning the post-pandemic city. As such, they are important for instigating new, inclusive, and even radical socio-technical paradigms for city-making and development in post-crisis contexts. City planners and elites ought not denigrate these as less developed and less sophisticated, divergent and discrepant, or lacking in their capacity to supplement and coexist with more modern, hegemonic approaches (Guma, 2022b; 2022c) as they do in fact characterize life across different spheres of heterogeneous urban contexts. When viewed through Alex and Mariam, the story of urban and infrastructural configurations during COVID restrictions in Nairobi is embedded by the subaltern experience and the limits of an exclusive and ineffective system or leaders and go-betweens that do not capture residents. It is important to envision a city where its plans are based on more heterogeneous approaches to standard and incremental technological infrastructures. This serves as a reminder of how urban populations live beyond the ideals and designs of networks.

Moreover, these stories highlight how residents seek to counter the deficits and anomalies of absence, lack, and incompleteness in formal structures that in an ideal world would be the custodians of the public good. Their strategies, practices, and forms highlight urban approaches that do not limit themselves to standard solutions for a technological fix or magic bullet in search of what works (Guma and Monstadt, 2021; Guma, 2022b). They reflect the need for research that attends to alternative reassemblages, such as modes that seek to reproduce life from outside of the established modes (ibid.). They offer a contextualized understanding of innovative and adaptive strategies developed by residents within the confines of their homes, neighbourhoods, and cities in order to cope with the adverse impacts and restrictive measures of the state. What these demonstrate is the role and impact of lockdown on the urban poor and their responses and recovery efforts, ranging from the use of digital technologies and survival infrastructures, to critical social modes of care and resilience in the face of pandemic and related crises. This is imperative given the need to build a better understanding of a range of institutions and structures that draw from different logics and considerations beyond centralized structures and outlooks.

Conceptually, the stories of Alex and Mariam offer a reading of infrastructure as relational, and infrastructure development as a process that is not affected solely by neoliberal interventions, but also by situated socio-material practices (see Guma, 2022a; 2022c). These open up further room for novel ways of seeing infrastructure in cities beyond conventional and completist frames, and draw direct attention to realities that may not always be clearly neat but uncertain, and entailing complex, contingent and heterogeneous elements and effects. For this reading, postcolonial discourses on cities of the Global South and African cities in particular provide latitude for achieving this objective. My goal in this chapter was to add to such work in further theorizing platform work, everyday life, and survival in contexts of urban heterogeneity. More research is needed to examine not just the bodies, claims, and struggles of platform work in hard times, but also the governance of such work and its effectiveness, inclusiveness, legitimacy, and significance in enhancing pathways for sustainable post-pandemic futures.

References

Amin, A. (2014) 'Lively infrastructure', *Theory, Culture & Society*, 31(7–8): 137–61.

Cardon, D. (2005) 'Innovation par l'usage', in A. Ambrosi, V. Peugeot and D. Pimienta (eds) *Enjeux de mots: Regards multiculturels sur les sociétés de l'information*, Caen: C&F Éditions.

Chéneau-Loquay, A. (2010) 'Innovative ways of appropriating mobile telephony in Africa', report published by the French Ministry of Foreign and European Affairs and the International Telecommunication Union (ITU), [online] September, available from: https://www.itu.int/ITU-D/cyb/app/docs/itu-maee-mobile-innovation-africa-e.pdf

Doherty, J. (2017) 'Life (and limb) in the fast-lane: Disposable people as infrastructure in Kampala's boda boda industry', *Critical African Studies*, 9(2): 192–209.

Dourish, P. and Bell, G. (2011) *Divining a Digital Future: Mess and Mythology in Ubiquitous Computing*, Cambridge, MA: MIT Press.

Edward, W. (2020) 'The Uberisation of work: The challenge of regulating platform capitalism. A commentary', *International Review of Applied Economics*, 34(4): 512–21.

Furlong, K. (2014) 'STS beyond the 'modern infrastructure ideal': Extending theory by engaging with infrastructure challenges in the South', *Technology in Society*, 38: 139–47.

Guma, P.K. (2022a) 'Thinking through frugal regimes of survival and improvisation in times of crisis: Considerations for urban research', *Zeitschrift für Ethnologie/Journal of Social and Cultural Anthropology*, 147(1–2): 113–18.

Guma, P.K. (2022b) 'Nairobi's rise as a digital platform hub', *Current History*, 121(835): 184–89.

Guma, P.K. (2022c) 'The temporal incompleteness of infrastructure and the urban', *Journal of Urban Technology*, 29(1): 59–67.

Guma, P.K. and Monstadt, J. (2021) 'Smart city making? The spread of ICT-driven plans and infrastructures in Nairobi', *Urban Geography*, 42(3): 360–81.

Guma, P.K. and Mwaura, M. (2021) 'Infrastructural configurations of mobile telephony in urban Africa: Vignettes from Buru Buru, Nairobi', *Journal of Eastern African Studies*, 15(4): 527–45.

Kenya National Bureau of Statistics. (2019) 'Kenya population and housing census – Economic Survey 2019', Nairobi: Government of Kenya.

Lee, A., Mackenzie, A., Smith, G.J.D. and Box, P. (2020) 'Mapping platform urbanism: Charting the nuance of the platform pivot', *Urban Planning*, 5(1): 116–28.

Olima, W.H.A. (2001) 'The dynamics and implications of sustaining urban spatial segregation in Kenya: Experiences from Nairobi Metropolis', Paper presented at International Seminar on Segregation in the City, 26–28 July, Nairobi: University of Nairobi.

Olvera, L.D., Plat, D. and Pochet, P. (2020) 'Looking for the obvious: Motorcycle taxi services in sub-Saharan African cities', *Journal of Transport Geography*, 88. https://doi.org/10.1016/j.jtrangeo.2019.102476

Scheepers, C.B. and Bogie J. (2020) 'Uber Sub-Saharan Africa: Contextual leadership for sustainable business model innovation during COVID-19', *Emerald Emerging Markets Case Studies*, 10(3): 1–18.

Sheller M. (2004) 'Mobile publics: Beyond the network perspective', *Environment and Planning D: Society and Space*, 22(1): 39–52.

Silver, J. (2014) 'Incremental infrastructures: Material improvisation and social collaboration across post-colonial Accra', *Urban Geography,* 35(6): 788–804.

Simone, A.M. (2004) 'People as infrastructure: Intersecting fragments in Johannesburg', *Public Culture*, 16(3): 407–29.

Simone, A.M. (2008) 'The politics of the possible: Making urban life in Phnom Penh', *Singapore Journal of Tropical Geography,* 29(2): 186–204.

Simone, A.M. (2016) 'The Uninhabitable? In between collapsed yet still rigid distinctions', *Cultural Politics*, 12(2): 135–54.

Sitas, R., Cirolia, L.R., Pollio, A., Sebarenzi, A.G., Guma, P.K. and Rajashekar, A. (2022) 'Platform politics and silicon savannahs: The rise of on-demand logistics and mobility in Nairobi and Kigali', Cape Town: African Centre for Cities, University of Cape Town.

Steen, M. (2011) 'Upon opening the black box of participatory design and finding it filled with ethics', Research paper for Nordic Design Research Conference 'Making Design Matter', Aalto University, Helsinki, 29–31 May 2011. https://archive.nordes.org/index.php/n13/article/downl oad/95/79

10

An Urban Data Politics of Scale: Lessons from South Africa

Jonathan Cinnamon

Introduction

In an era defined not so much by the rapid increase in data availability but by the expectation that data will transform society (Markham, 2013; van Dijck, 2014; Srinivasan et al, 2017), cities have become a key site for speculative investment in data (Barns, 2018; Das, 2020). Many city governments around the world now subscribe to the idea that myriad challenges – from climate change to traffic congestion and economic competitiveness – can be effectively addressed through data practices at the urban scale. The notion of 'smart cities' captures this line of reasoning, the belief that real time sensor measurements, computational forms of knowledge production, and a mode of action-oriented governance motivated by creativity and experimentalism will guide future cities into an optimized version of the present. Similarly, ambitions towards data intensive modes of action are also prevalent across the private sector and, more recently, within civil society, non-governmental, and grassroots organizations. Although their motivations may vary, 'data-driven' has become a near universal mantra of diverse urban actors, providing a powerful basis for critically examining the implications and outcomes of data investments.

While the critical smart cities literature has drawn substantial attention to a growing mode of governance driven by data and technology (Datta and Odendaal, 2019; Guma and Monstadt, 2021; Sadowski, 2021; Söderström et al, 2021), less critical attention has been focused on the implications when non-governmental organizations seek to position themselves as data-driven actors. Focusing at the grassroots level, this analysis inhabits a particular moment in South African cities in which 'data' emerged as a powerful trope within civil society organizations, and social movements. In the early 2010s, as municipal governments in the country's metropolitan areas began chasing globally circulating ideas of data-driven policy making and innovation (see Blake et al, Chapter 11, this volume), local community actors also became caught up in the opportunities promised through data-driven evidence. In turning to data-driven forms of social activism, this period marked a notable shift in the discursive and representational tactics of many civil society organizations and social movements in South African cities.

South African grassroots organizations and social movements have a long history of tactical manoeuvring to empower actors involved in localized struggle, including shifting alliances, altering discourses, and reorienting to new spaces of action (Miraftab, 2020). Under such conditions, South African grassroots groups have traditionally deployed a variety of spatial tactics to create new political terrains, in a country where spatiality, rather literally, shapes social relations (Andres et al, 2020). Space provides an obvious background for social struggle when social inequity is carved into the very fabric of the South African city, as both the legacy of apartheid planning and a visible reminder of the inadequacies of social redress efforts in an era of democracy and neoliberal urbanization (Massey and Gunter, 2020; Smith, 2022). In particular, grassroots organizations leverage *scalar* tactics – representations and discourses of spatial scale – to make visible the vast social inequalities in South African cities. For grassroots actors, representations and discourses of localized hardship provide a powerful, visceral means to disrupt 'global city' (McDonald, 2008) narratives authored by governments and those in power. In doing so, activists enact a politics of scale in which scalar relations are weaponized as a means to achieve political goals. Vanessa Watson's (2003) concept of 'conflicting rationalities' captures the fundamental divisions in many Global South cities between techno-managerial elites and highly marginalized urban populations (De Satgé and Watson, 2018; Guma et al, 2022). For Ngwenya and Cirolia (2021, 691, emphasis added), the concept 'provides insights into the *seemingly irreconcilable world views of different actors* involved in complex development processes'. Here, as I will show, irreconcilable world views in Cape Town play out in scalar terms, as a politics of scale fuelled by opposing global and local frames. For the City of Cape Town (hereafter the City), a manifestly global worldview comes into direct and ongoing conflict with the fundamentally local perspective of social movement actors. Because of this, scalar contestations pervade the urban political condition in Cape Town. Tracing this scalar conflict through a recent phase of data investments at both the city and grassroots level, this chapter critically examines the ability of data-driven tactics to drive an overall politics of scale.

In the context of an ongoing service delivery crisis in urban South Africa, this chapter examines what happens when the politics of urban data meets the politics of scale in a Cape Town context defined by conflicting rationalities and opposing scalar frames. More specifically, it examines the consequences when new data-driven tactics of auditing and counting are pursued by grassroots actors to make the crisis visible, set against the longer history of antagonistic actions against injustice and inequality in the South African context. Previously, I have analysed grassroots data activism through the lens of data power (Cinnamon, 2020a), revealing how an over-reliance on data served to disempower social movements. Here, I add to this analysis by focusing on the ways in which divergent actors – in government and in grassroots social movements – deploy scalar frames in urban data politics. The following section provides a review of the concept of geographic scale, focusing on how scalar discourses and representations are weaponized in a politics of scale. The main section shows how, under conditions of contested rationalities symbolized by fundamentally different scalar imaginaries of Cape Town, grassroots actors embraced data politics in the struggle for dignity in sanitation in the city's informal settlements. In revealing the limitations of data and a subsequent remobilization of scalar tactics, the final section links data at the grassroots level with the post-political urban condition, suggesting a need to consider what forms of politics data enables and what forms it forecloses.

Social movements and the politics of scale

Geographic scale is central to the geographic imagination, and it also pervades discourses in wider public and policy spheres. Sometimes controversially understood as a geographic 'master concept' superseding other forms of spatiality (such as place, region, landscape), scale nonetheless provides a powerful vocabulary and conceptual frame for articulating the complexity of the social world (Howitt, 1998). Yet scale itself is expressed and deployed in varied ways both by geographers and in the wider public arena. When understood as a material entity, scale is frequently articulated using a vertical or hierarchical 'levels' ontology (MacKinnon, 2011) comprised of horizontally bounded and discrete spaces in which daily life and social processes occur, variously described as 'platforms' (Smith, 2004), 'containers' (Moore, 2008), 'arenas' (Swyngedouw, 2004), 'or 'spatial units' (Brenner, 2000). Scale is also conceptualized epistemologically as a representational or discursive frame (Blakey, 2021). Across the range of material metaphors and epistemological constructions, scale provides a means to examine social processes as phenomena operating *within* spaces described as global, national, regional, urban, community, and home, as well as the personal scale of individual bodies. In all conceptualizations, however, scalar categories perform an ordering function, reproducing binaries such as local/global and micro/macro (Mountz and Hyndman, 2006). Demonstrating the falsity of such dualisms, relational understandings of scale have taken prominence in academic geography, providing a means of understanding how social processes are produced both *within* and *across* scales – from the macro of the global to the micro of the intimate (Brenner, 1998; Howitt, 1998; Mountz and Hyndman, 2006; Pratt and Rosner, 2006; McCann and Ward, 2010).

Defining and deploying scale in geographic research is, therefore, a highly contested topic of debate for at least two key reasons: (1) the variability with which it has been used in academic and public spheres; and (2) the notion that scalar frames act to delimit how we can know the social world. Regarding the latter, when predefined scalar categories are superimposed over the social world, they provide an actionable framework for analysis, but in doing so they choke off the possibility of other ways of knowing social relations (Blakey, 2021). Critics have long argued that scalar terms and categories must be understood as socially constructed rather than already existing 'givens' out there in the world (Smith, 2003); yet once enrolled into language and representation, they become reified as 'real' material things. As Howitt (1998, 50) explains, scalar terms have 'lost their identity as analytical abstractions and have come to be seen as things in themselves to be dealt with categorically'. This matters because once constructed, scalar frames performatively shape real-world social, economic, and political processes and outcomes (Delaney and Leitner, 1997). Thus, for some critics, scale is too problematic and must be expunged from our theoretical and analytical vocabularies (Marston et al, 2005).

Yet scale remains a central component of wider geographic imaginaries; it is a 'fictitious reality' coursing through academic, public, and policy discourses (Smith, 2003). For examining social movements and grassroots activism, dropping scale from our analyses is a particularly tall order, given the normative scalar sensibilities invoked by terms such as 'community-based', 'ground-level', and 'grassroots'. And for Leitner and Miller (2007), scale is a crucial instrument of leverage for social movements; under the shifting politics of neoliberalism, social movements seek out multiscalar strategies, form translocal alliances, and explicitly debate at what scale organizing should take place. For Moore (2008), scale is appropriate when treated not as a *category of analysis* – whereby researchers impose scalar categories on their analysis of social processes – but rather, as a *category of practice* that real-world actors actually use to advance their goals. In other words, researchers should examine the tendency of various social actors 'to partition the social world into hierarchically ordered spatial "containers"' (Moore, 2008, 212) rather than attempting to use these 'spatial containers' as lenses to frame our understand of the world. This standpoint asks us to resist the normative urge to a priori apply a scalar frame, but rather to examine how scale emerges and is deployed as a category of practice by various actors. Such an approach 'focuses attention on the scalar classifications and discourses deployed by different political actors and movements' in order to legitimize and empower their actions (MacKinnon, 2011, 29).

From this perspective, based on a pragmatic understanding of scale as central to wider geographic imaginaries, researchers have examined how actors engage in scalar actions when negotiating power relations (Pesqueira and Glasbergen, 2013). Understanding scale as more explicitly *politically* constructed (Delaney and Leitner, 1997; Brenner, 1998), the 'politics of scale' refers to 'how scales and scalar relations are shaped by the processes of struggle between powerful social actors and subaltern groups' (MacKinnon, 2011, 24). Research on the politics of scale builds from the scale debates to examine how scale is used to gain strategic advantage, influence how social and political issues are understood, and as a discursive tool for control over social and political space (Brenner, 1998). When used as part of a politics of struggle, scale becomes a device that can be weaponized against competing actors through the production of alternative scalar frames (Kurtz, 2003). As Neil Smith (1992, 78, emphasis added) explains, '[b]y setting boundaries, scale can be constructed as a means of constraint and exclusion, a means of imposing identity, but *a politics of scale can also become a weapon of expansion and inclusion*, a means of enlarging identities'. For traditionally disempowered actors, a variety of scalar tactics are weaponized as tools of expansion and inclusion. Tactics of 'scale bending' attempt to challenge entrenched ideas about appropriate scales of action (Smith, 2004). Such tactics can be used by grassroots groups to, for instance, contest central government inaction concerning injustice at the local scale. Tactics of 'scale jumping' are used to elevate struggles to larger and more visible registers in the scalar hierarchy, such as city, national, or global to align a local movement with broader solidarities (MacKinnon, 2011; Bond and Ruiters, 2017). Similarly, 'multiscalar' tactics are used to connect issues at one scale with processes occurring at another scale (Leitner et al, 2008). These attempts to variably 'rescale', 'fix', or 'undo' scale are often obligatory in many forms of grassroots struggle. As Kevin Cox (1998) explains, advancing local causes often requires taking the struggle out of the locally bounded territory – its 'space of dependence' – to a 'space of engagement', typically a larger spatial scale at which the struggle can be made visible and actionable.

Urban data and the politics of scale in Cape Town

Against the twin backdrops of data-driven urban actors and the politics of scale, this section considers what happens when data and scale meet in the context of social movement activism. Specifically, the analysis examines a brief period in time within a longer struggle against spatial injustice when data-driven tactics momentarily took centre stage in South African social movements. Grassroots actors in South Africa often shift tactics to advance their goals because, as Miraftab (2020, 437) explains, success requires an 'agility and ability to transgress and destabilize hegemonic normality and open up new political terrains for the imagination of a different future'. An orientation to data-driven tactics took place amidst the larger 'data turn' of the 2010s, which saw cities in South Africa and around the world pursue ambitions of data-driven policy making and innovation.

Over the following two subsections, this analysis will show how activists' embrace of data at the *tactical* level came at the expense of a broader scalar politics at the level of the *strategic*. Here, a distinction between tactics and strategies provides a way of understanding how everyday discourses and representations are leveraged by different actors in order to advance their overall political objectives. For de Certeau, tactics are understood as specific, everyday actions of resistance, while strategies exist on a higher plane, are longer-term in orientation, and are available only to powerful organizations such as governments (de Certeau, 1984). Below I outline the centrality of scale in both governance and grassroots politics in South Africa, and then critically examine how investing in data-driven tactics against the ongoing crisis of indignity in sanitation had implications for the wider strategic goals of the social movement.

Scalar politics in governance and the grassroots

To understand the power of the politics of scale for grassroots organizations in urban South Africa, it is instructive to examine how scalar discourses and representations permeate the governance sphere, and the influence this has at the grassroots. As a form of spatiality widely deployed in practice (Moore, 2008), scalar politics are pervasive at the municipal government level, particularly in Cape Town. As a 'strategy-led organization' (de Lille and Keeson, 2017), in recent years the City has tried to bypass national-level administration on issues of global affairs and strategically distance itself from other South African cities. Cape Town was an originating member of the South African Cities Network (SACN) when it was formed in 2002, a coalition between the country's major metropolitan areas that 'encourages the exchange of information, experience and best practice on urban development and city management' (South African Cities Network, 2022). However, the City later ended its formal affiliation with the network, and since then has more actively tried to reposition itself on the global stage, as a member of a more nebulous but exclusive network of 'global cities'. Patricia de Lille, the mayor of Cape Town from 2011 to 2018, explains the motivations in a book reflecting on her experience leading the city (de Lille and Keeson, 2017):

Our aspiration is to make Cape Town a truly world-class city. This means focusing on key partnerships and connections, as well as understanding how we can promote the city effectively. It means focusing our international agenda on relationships and initiatives that can create real value for the people of Cape Town [...] Thinking about Cape Town without thinking about how it fits into the world would be a serious failure of vision and leadership.

Elsewhere, former mayor de Lille explicitly revealed a desire to jump scales:

We try to put the bar very high and not compete with Johannesburg and Durban. We live in a global village. And to be competitive you must compete with cities like Singapore, Vancouver, New York and Sydney.
(The Worldfolio, 2014)

Of particular relevance to the scale-jumping ambition, the City has been keen to invest in digital and data-led initiatives to fuel the city's global competitiveness, as part of their flagship 'Digital City Strategy'. Developed in collaboration with international partners in the United States and Europe, a key overall aim of the strategy was to elevate the city's status in the scalar hierarchy, positioning Cape Town as 'Africa's Digital City' – the continent's main node in the global network of smart cities, and *the* gateway to Africa for technology-related economic investment (Cinnamon, 2022). On the occasion of the release of the City's open data portal in 2015 – a legacy of the city's 2014 tenure as 'World Design Capital' – former mayor de Lille explains how data initiatives can drive their scale-jumping ambitions:

This is a historic day, as we join cities such as New York, London and Helsinki that have forged the way for cities to make their data sets available to the public. In today's knowledge economy, access to data is instrumental in becoming competitive.
(PoliticsWeb, 2015)

Paralleling – and to a degree imitating – the City's scalar politics, grassroots actors in Cape Town also use scalar discourses and representations to advance their ambitions. As part of a more widely held spatialized understanding of South African society, grassroots actors emphasize a spatialized politics because apartheid-era geographies remain a defining feature of the urban landscape in the democratic era. Justice then, is necessarily a geographical concept in South Africa, and South African grassroots groups frequently engage in spatial and scalar tactics. For activists and civil society organizations in Cape Town, their modus operandi often requires explicitly challenging the 'Cape Town as global city' imaginary, a partial and incomplete spatial representation that serves to invisibilize the city's vast geographies of poverty and exclusion, represented in particular by informal settlements. In the context of contesting spatial injustice in a setting historically defined by conflicting rationalities (Watson, 2003), a politics of scale has long played out tactically. More recently, however, in an attempt to access the level of the strategic enjoyed by governments and other elite actors, a diverse range of activists, civil society organizations, and social movements have banded together to act as the de facto authority of the country's informal settlements under conditions of service delivery crisis due to state neglect.

Drawing on a longer history of anti-apartheid resistance, and fuelled by opposition to the neoliberalization of post-apartheid socio-economic transformation, a maturing collaborative model of 'alliances' and 'networks' has strengthened the power of civil society and social movements in contemporary South Africa (Ballard et al, 2006; Madlingozi, 2007). Although specific causes vary, from equity in the delivery of basic services to the location of low-income housing, social movements are emerging as the quasi-official voice of the country's informal spaces, in contrast to the government's territorial focus on elite spaces. Following de Certeau (1984, xix), I argue that social movements and civil society have thus gained access to the plane of the strategic, since doing so requires 'a place that can be circumscribed [...] and thus serve as the basis for generating relations with an exterior distinct from it'. For the increasingly powerful social movement networks, elevating local causes to a higher plane of authority is the overall strategy. And mimicking the strategy of the city government, this necessarily requires an explicitly scalar approach in a country where achieving social and political recognition is contingent on making local causes known to the state (Parnell and Pieterse, 2010; Levenson, 2021).

Scale, data, and the movement for dignity in sanitation

The ongoing service delivery crisis and struggle for dignity in sanitation in informal settlements illustrate the overall scale-jumping strategy of social movements in South Africa, and how data was mobilized to advance this strategy. As part of the Ses'khona People's Rights Movement, the emergence of a widespread, highly networked sanitation movement marked a period in South African cities defined by overt antagonism and opposition to intransigent and indeed widening social inequalities (Baxter and Mtshali, 2020; Jabary Salamanca and Silver, 2022). By 2010 – the year the FIFA World Cup was held across the country – Cape Town was experiencing all-out 'toilet wars' (Robins, 2014a) as the struggle for sanitation gathered strength to contest the city's inability or unwillingness to provide sanitation services in informal settlements (McFarlane and Silver, 2017; Ernsten, 2019). Key to the struggle was the use of spatial tactics, including deploying discourses and representations of the local as a weapon to draw attention to the unsanitary conditions in informal settlements. Most spectacularly, activists removed human excrement from informal settlements and deposited it in areas of the city emblematic of the 'Cape Town as global city' imaginary, including the N2 motorway, Cape Town International Airport, and the provincial legislature. As Robins (2014b, 1) puts it, '[b]y taking their struggle to global sites of tourism [...] sanitation activists had raised the stakes in the ongoing politicization of shit in the Western Cape [Province]'. In other scalar – yet somewhat less spectacular – tactics organized by allies in the Social Justice Coalition (SJC) and other grassroots partners, thousands of residents were transported out of informal settlements to queue up for the public toilet facilities in the city centre, and 2,500 went to the very peak of municipal administrative power, former mayor de Lille's office, to protest (McFarlane and Silver, 2017). From a politics of scale lens, I suggest that activists used multiscalar tactics based on a relational understanding of scale, which 'simultaneously [broadens] the scale of action while drawing strength from reinforcing the local scale' (Leitner et al, 2008, 160). Indeed, in attempting to elevate the local, activists do not reject the 'global city' representation at all – instead, they use it to their advantage based not on an 'either/or' but a 'both/and' understanding of spatial scale (Harvey, 2000).

For activists, relational multiscalar tactics reveal how the city exists at the intersection of highly divergent global and local realities; it is a city defined by conflicting rationalities produced, in part, at the place where the local and the global meet. Such relational multiscalar tactics serve to elevate the status of the local scale in how Cape Town is imagined, a scalar frame literally relegated to the margins of the City's prevailing global city scalar imaginary. Prior to these grassroots scalar manoeuvres the crisis of sanitation in informal settlements was largely invisible, absent from media and political discourse. Their success lies in making sanitation visible to those in power. As Odendaal (2019, 170) explained in this context, '[t]he power of the spectacle lies in elevating issues to policy discourses'.

Multiscalar tactics brought material success in the struggle for sanitation, notably the introduction of a City-supplied janitorial service to clean and maintain toilets, first promised in 2011. A press release by the SJC at the time praised the mayor and the City for their 'significant shift' in sanitation policy. At the same time, it also foretold of an imminent shift in how the sanitation movement positioned itself. In line with their growing reputation as the voice of informal settlements, the SJC and their partners in the sanitation movement began to discursively recast their status, from grassroots agitators to an empowered coalition of informal settlement advocates operating at the same level as government. As stated in the press release (Social Justice Coalition, 2011):

The SJC and its partners – including community forums, social movements, ward councilors, NGOs, faith based organisations, technical experts, and academic institutions – are committed to working with the City to ensure that over time, every person has access to a toilet and water source that is clean, safe and dignified. Janitorial services would be notable stride towards this objective, and would serve to illustrate the importance of fostering constructive partnership between government and communities.

Moreover, the 2011 press release also foretold of a shift in the tactics deployed by the SJC and their social movement allies. In subsequent years they would set aside multiscalar tactics – which brought the realities of local informal settlement life to the globalized spaces of Cape Town – to more fully invest in tactics of evidence-based accountability (Social Justice Coalition, 2011):

> While the Mayor's commitment is significant, it must now be developed into a workable implementation plan with a clear timeline for a service that is both practical and accountable. *Such a plan will require detailed discussion about how communal toilets are distributed, monitored, and maintained* [emphasis added].

While the scalar *strategy* remained the same – elevating local grassroots issues through jumping scales – this quote marks the beginning of a shift from *scalar tactics* to *data tactics*, from antagonism to accounting. By 2013, the SJC and partners had declared the 'Power of Data as Evidence' (Russell, 2013) and had conducted the first of several 'social audits' on sanitation services in informal settlements to address their concern that the City's new janitorial service must be monitored (Social Justice Coalition, 2013). Akin to a formal audit of a business, social audits are a tool for local communities and social movement partners to monitor government expenditures and delivery of services from the external vantage point of those set to receive the goods or services (Social Justice Coalition et al, 2015). One of the key functions of social audits is to compare the budgeted provision of services and infrastructure with the reality of service provision at the ground level through systematic collection and analysis of data. Although local residents and their partners in civil society inherently know that inequities in service provision are centred on informal settlements, the data-led approach was pursued in order to develop clear evidence of service gaps, data that could be used tactically as incontrovertible proof of service delivery injustice. Across a range of social audits conducted in Cape Town and other jurisdictions, the findings clearly show how municipal governments and their contractors had not fulfilled their service provision duties as laid out in budgeting and expenditure documents. In the turn away from scalar tactics towards data-driven tactics of monitoring and accountability, data becomes the weapon, as explained in the SJC's 'Power of Data as Evidence' document:

> It is easy to complain and say there are not enough toilets. And while anecdotal reports of sanitation issues are important, until you have the data – how many toilets there should be as opposed to how many there are, how many are being paid for with taxpayer money, and how many people have to use them – *you do not have the ammunition to launch a fact-based protest.*
>
> (Russell, 2013, emphasis added)

Substantial data hype emerged as a key outcome of Cape Town's first sanitation social audit in 2013. And by 2016, around a dozen social audits had been undertaken across the country, the Social Audit Network (2016) was birthed as a grassroots alliance of social auditing organizations and allies across the country, and 'fact-based protest' had reached the social movement mainstream. Social audit reports are extensive, detailed, professionally produced, and on completion presented to government ministers as evidence of service need. In many social audits, presenting the final report was seen as a neutral act, an opportunity to assist the government in a way that was free from the oppositional politics of many other forms of protest. As one social auditor explained to me:

> 'What I love about social audits, the tool itself, it makes sure that all participants come together [...] and it helps the government to plan better because now they know what the problem is, what to prioritize, what not to prioritize.'
>
> (Interviewee, civil society organization)

Through social audits, the sanitation movement's early scalar tactics took a backseat to tactics of objectivity, in the material form of data, reports, spreadsheets, and statistics. Yet shifting the tactics from scale to data met with comparatively little success. In almost all cases, social audit findings were not only widely dismissed by governments citing issues of methodological and statistical validity, they were used as an opportunity to deflect responsibility back to social audit organizations and informal settlement residents themselves (Cinnamon, 2020a; 2020b).

A 2016 social audit on safety and sanitation in schools in Western Cape Province illustrates this move (Equal Education, 2016). The report highlighted concerning issues of sanitation and safety, yet it was initially completely ignored despite the extensive and detailed findings presented. Interestingly, when a response by the minister was finally received (actually it was posted publicly to PoliticsWeb, a South African news and politics website), it drew on a scalar frame to resist the social audit findings. First, the minister 'downscaled' the findings by arguing that the data has no validity outside of the local area:

> There are crucial errors contained in the methodology of this audit. By their very nature, social audits rely on convenience sampling and/or purposive sampling. Both sampling methods are not scientific and *there is no systemic way of applying the findings beyond the areas in which they are found.* Consequently such studies do not have the potential to inform policy.
> (Schäfer, 2016, emphasis added)

Second, the minister used the tactic of 'scale bending' (Smith, 2004) to challenge a key assumption made in the social audit – that the issues should be dealt with solely by the provincial ministry of education and not at other scales of governance:

> [it is] unfair to blame the education department for social issues that need to be dealt with across the board by *numerous role-players in national, provincial and local spheres of government. The role of parents must also be highlighted.* There are still far too many parents who neglect their parental responsibilities. Perhaps Equal Education could launch a parent responsibility campaign across the country to impress on parents the important role they play in creating the kind of culture that we all want to see.
> (Schäfer, 2016, emphasis added)

This deflective move aimed to *rescale responsibility* for safety and sanitation concerns in schools, dispersing it through the scalar hierarchy from the national scale down to the individual. In doing so it clearly demonstrates how – through discursive and representational tactics – scale is effectively wielded as a political weapon against adversaries.

Scalar returns

As the data decade progressed, early hype around social auditing dissipated as a number of social movement actors observed that focusing on data can serve to disconnect the tactics from the strategy, in effect disempowering the movement. As a key figure in the sanitation social audits explained to me, 'I would never go head-to-head with a government on data again because to me it's not what wins the battle. What wins the battle is better understanding power and your leverage over them' (interviewee, civil society organization). Reflecting on the limited success of the data-driven approach, a comment by a participant at an event in 2017 is illustrative of a transition away from data within some grassroots organizations:

'There are more adversarial political actions – and I think if the government is failing to respond in processes such as this social audit process – that we should maybe embark on as civil society organizations here.'

More adversarial political actions soon followed. Recent developments in the Reclaim the City movement in Cape Town suggests a reinvigoration of scalar tactics. Reclaim the City first emerged in 2016 to contest the ongoing crisis of land and housing injustice, which was initiated under apartheid, but continues to relegate poor and lower-income residents to marginal and distant areas of the city (Odendaal, 2019; Herold and DeBarros, 2020). The key aims of the movement are to force the City to allocate centrally located lands for affordable housing rather than sell it off to developers, and to contest the displacement of low-income residents from gentrifying areas. While data and evidence remain an important component of the struggle, the movement deploys a range of tactics; in the words of the civil society organization Ndifuna Ukwazi (a key partner in the Reclaim the City movement), '[p]ower adapts and activists and tactics must diversify to remain effective' (Lessons for Change, nd).

A politics of scale has been important to the Reclaim the City movement, in part, I would argue, due to its locational focus on the 'globalized' City centre instead of the periphery. By juxtaposing the needs of local low-income residents against the desires of wealthy transnational developers seeking to profit from the City's reputation as a global city, the campaign has had material success. At the tactical level, the movement has used similar multiscalar tactics as seen in the early phases of the sanitation social movement. In one example, homeless members of the movement conducted protests and wrote slogans in chalk on the street in affluent areas of the city (Lessons for Change, nd). In another important action, a vacant City-owned building – the former Woodstock Hospital, renamed Cissie Gool House in honour of a local anti-apartheid activist – was 'symbolically occupied' in 2017 to contest rapid gentrification and displacement of working-class people from Woodstock and other areas of the City (Arderne, 2022). Now home to around 1,200 people over five years later, the ongoing occupation symbolizes the Reclaim the City movement's localized fight against the forces of globalized capital (Dougan, 2018; Ngwenya and Cirolia, 2021), and more generally, the power of scalar tactics in a setting defined by contrasting scalar frames.

Urban data and the post-political?

In an editorial introduction to a journal special issue on scalar transformations in post-apartheid South Africa, Richard Ballard explains the significance of scale and its contestations; as the author explains, 'South Africa's transition to democracy has been accompanied by remarkable dynamism and experimentation with the scales at which spatial arrangements are configured' (Ballard et al, 2021, 137). In the context of experiments with data-driven tactics against the ongoing crisis of service delivery injustice, this chapter asked what happens when a data politics is asked to contribute to a politics of scale in a Cape Town context defined by conflicting rationalities and their opposing scalar frames. As laid out in detail in this chapter, one possible answer to this question is that data serves to disempower a scalar politics when data tactics are used to fuel an overall scale-jumping strategy.

In the context of contentious politics, scale-jumping provides social movements with a way to overcome the 'limitations of localness' (Leitner et al, 2008, 159–160; Herrera, 2022). Yet a key claim made here is that the tactics used to advance this overall strategy matter. Elsewhere, I examined how the embrace of data-led social audits marked a shift away from 'people power' activism to a form of activism driven by unrealized imaginaries of 'data power', which had consequences for activists' ability to advance their political goals (Cinnamon, 2020a). In that piece I conceptualized data power as 'relational, partial, and provisional, and enacted through the co-constitution of people, technologies, and political discourses' (Cinnamon, 2020a, 636). The present analysis extends that conceptualization to reveal the additional power of spatial scale in a setting dominated by scalar frames in both government and grassroots discourse. But rather than adding an additional dimension of 'scale power' to a people power/data power binary, the analysis presented in this chapter suggests that scale – like spatiality more broadly – is something more interwoven and cross-cutting when space is the very basis of contentious politics. Early successes in the sanitation movement reveal the absolute necessity for, and efficacy of, a politics of scale under conditions of conflicting rationalities; a politics of scale provided a means of making injustice visible at larger scalar registers, and it enabled activists to contest how governments pursue global ideas while distancing themselves from the local reality of service delivery crises. In the context of an ongoing struggle for dignity in sanitation, a turn to data tactics and a retreat from relational multiscalar tactics can be understood as an attempt to strategically position the movement on the same level as government, as their accountants rather than their antagonists. The limited success of this approach provides a cautionary tale for future social movement activisms in search of new weapons, in Cape Town and elsewhere. For the wider domain of urban data politics, it suggests a need to consider what forms of politics data enables, and what forms of politics data forecloses.

These findings link the use of data as a grassroots tactic to the foreclosure of agonistic forms of political interaction and, potentially, the *depoliticization* of social movements. Via the lens of post-politics, depoliticization involves two processes: (1) consensus-building and the marginalization of dissent; and

(2) the reduction of complex societal issues to mere technical challenges that can be solved through technology rather than through a more substantive reorganization of economic or political arrangements (da Schio and van Heur, 2022). As this analysis demonstrated, the embrace of data-driven tactics at the grassroots engages both of these processes. In another analysis (Cinnamon, 2020b) I examined how grassroots actors can strategically engage with policy makers to advance their political goals, showing how technical and consensus-driven interactions can be advantageous when the aims are instrumental (achieving specific tangible changes). Yet such engagements may not be appropriate when the aims are normative (challenging deep-rooted power asymmetries). In a Cape Town context of intractable crises, conflicting rationalities, and opposing local and global scalar frames, data-as-tactic steered social movements towards the former aim at the expense of than the latter, without achieving substantive tangible change. I would argue that the power of multiscalar tactics lies in their relationality, their ability to transgress categories and *produce* both space and social possibilities anew (Lefèbvre, 1991; Herrera, 2022). In contrast, data tactics, based on categorical understandings of the world, are often limited ontologically to the mere representation of the world via already existing spatial frames. Thus, for social movements seeking to fundamentally disrupt the scalar and social order, investing in data-driven tactics might well serve to constrain possibilities to the spatial status quo.

References

Andres, L., Jones, P., Denoon-Stevens, S.P. and Lorena M. (2020) 'Negotiating polyvocal strategies: Re-reading de Certeau through the lens of urban planning in South Africa', *Urban Studies*, 57(12): 2440–55.

Arderne, M. (2022) 'Residents, not occupiers, live at Cissie Gool House', New Frame, [online] 2 March, available from: https://www.newframe.com/residents-not-occupiers-live-at-cissie-gool-house/

Ballard, R., Habib, A. and Valodia, I. (eds) (2006) *Voices of Protest: Social Movements in Post-Apartheid South Africa*. Durban: University of KwaZulu-Natal Press.

Ballard, R., Parker, A., Butcher, S.C., de Kadt, J., Hamann, C., Joseph, K., Mapukata, S., Mkhize, T., Mosiane, N. and Spiropoulos, L. (2021) 'Scale of belonging: Gauteng 30 years after the repeal of the Group Areas Act', *Urban Forum*, 32(2): 131–39.

Barns, S. (2018) 'Smart cities and urban data platforms: Designing interfaces for smart governance', *City, Culture and Society*, 12: 5–12. https://doi.org/10.1016/j.ccs.2017.09.006

Baxter, V. and Mtshali, M.N. (2020) '"This shit is political; shit is real." The politics of sanitation, protest, and the neoliberal, post-apartheid city', *Studies in Theatre and Performance*, 40(1): 70–82.

Blakey, J. (2021) 'The politics of scale through Rancière', *Progress in Human Geography*, 45(4): 623–40.

Bond, P. and Ruiters, G. (2017) 'Uneven development and scale politics in southern Africa: What we learn from Neil Smith', *Antipode*, 49(S1): 171–89.

Brenner, N. (1998) 'Between fixity and motion: Accumulation, territorial organization and the historical geography of spatial scales', *Environment and Planning D: Society and Space*, 16(4): 459–81.

Brenner, N. (2000) 'The urban question: reflections on Henri Lefebvre, urban theory and the politics of scale', *International Journal of Urban and Regional Research*, 24(2): 361–78.

Cinnamon, J. (2020a) 'Attack the data: Agency, power, and technopolitics in South African data activism', *Annals of the American Association of Geographers*, 110(3): 623–39.

Cinnamon, J. (2020b) 'Power in numbers/Power and numbers: Gentle data activism as strategic collaboration', *Area*. https://doi.org/10.1111/area.12622

Cinnamon, J. (2022) 'On data cultures and the prehistories of smart urbanism in "Africa's Digital City"', *Urban Geography*, 44(5): 850–70. https://doi.org/10.1080/02723638.2022.2049096

Cox, K.R. (1998) 'Spaces of dependence, spaces of engagement and the politics of scale, or: looking for local politics', *Political Geography*, 17(1): 1–23.

da Schio, N. and van Heur, B. (2022) 'Resistance is in the air: From post-politics to the politics of expertise', *Environment and Planning C: Politics and Space*, 40(3): 592–610.

Das, D. (2020) 'In pursuit of being smart? A critical analysis of India's smart cities endeavor', *Urban Geography*, 41(1): 55–78.

Datta, A. and Odendaal, N. (2019) 'Smart cities and the banality of power', *Environment and Planning D: Society and Space*, 37(3): 387–92.

de Certeau, M. (1984) *The Practice of Everyday Life*, translated by S. Rendall, Los Angeles: The University of California Press.

de Lille, P. and Keeson, C. (2017) *View from City Hall: Reflections on Governing Cape Town*, Johannesburg and Cape Town: Jonathan Ball.

de Satgé, R. and Watson, V. (2018) *Urban Planning in the Global South: Conflicting Rationalities in Contested Urban Space*, Cham: Palgrave Macmillan.

Delaney, D. and Leitner, H. (1997) 'The political construction of scale', *Political Geography*, 16(2): 93–7.

Dougan, L. (2018) 'Gentrification victims: The Dreyer family's struggle to keep a roof over their head under a city bridge', Daily Maverick, [online] 17 September, available from: https://www.dailymaverick.co.za/article/2018-09-17-gentrification-victims-the-dreyer-familys-struggle-to-keep-a-roof-over-their-head-under-a-city-bridge/

Equal Education. (2016) 'Of "loose papers and vague allegations": A social audit report on the safety and sanitation crisis in Western Cape schools', Cape Town: Equal Education, available from: https://equaleducation.org.za/wp-content/uploads/2016/09/Western-Cape-Schools-Safety-and-Sanitation-Social-Audit-Report.pdf

Ernsten, C. (2019) 'Utopia and dystopia in the post-apartheid city: The praxis of the future in Cape Town', *Social Dynamics*, 45(2): 286–302.

Guma, P.K. and Monstadt, J. (2021) 'Smart city making? The spread of ICT-driven plans and infrastructures in Nairobi.' *Urban Geography,* 42(3): 360–81.

Guma, P.K., Monstadt, J. and Schramm, S. (2022) 'Post-, pre- and non-payment: Conflicting rationalities in the digitalisation of energy access in Kibera, Nairobi', *Digital Geography and Society*, 3. https://doi.org/10.1016/j.diggeo.2022.100037

Harvey, D. (2000) *Spaces of Hope*. Los Angeles: University of California Press.

Herold, B. and DeBarros, M. (2020) '"It's not just an occupation, it's our home!" The politics of everyday life in a long-term occupation in Cape Town and their effects on movement development', *Interface: A Journal for and about Social Movements*, 12(2): 121–56.

Herrera, J. (2022) *Cartographic Memory: Social Movement Activism and the Production of Space*, Durham, NC: Duke University Press.

Howitt, R. (1998) 'Scale as relation: Musical metaphors of geographical scale', *Area*, 30(1): 49–58.

Jabary Salamanca, O. and Silver, J. (2022) 'In the excess of splintering urbanism: The racialized political economy of infrastructure', *Journal of Urban Technology*, 29(1): 117–25.

Kurtz, H.E. (2003) 'Scale frames and counter-scale frames: Constructing the problem of environmental injustice', *Political Geography*, 22(8): 887–916.

Lefebvre, H. (1991) *The Production of Space*, translated by D. Nicholson-Smith, Cambridge, MA: Blackwell.

Leitner, H. and Miller, B. (2007) 'Scale and the limitations of ontological debate: A commentary on Marston, Jones and Woodward', *Transactions of the Institute of British Geographers*, 32(1): 116–25.

Leitner, H., Sheppard, E. and Sziarto, K.M. (2008) 'The spatialities of contentious politics', *Transactions of the Institute of British Geographers*, 33(2): 157–72.

Lessons for Change. (nd) 'Tactics – Be flexible', Lessons for Change, [online], available from: https://www.lessonsforchange.org/reclaim-the-city/

Levenson, Z. (2021) 'Becoming a population: Seeing the state, being seen by the state, and the politics of eviction in Cape Town', *Qualitative Sociology*, 44(3): 367–83.

MacKinnon, D. (2011) 'Reconstructing scale: Towards a new scalar politics', *Progress in Human Geography*, 35(1): 21–36.

Madlingozi, T. (2007) 'Post-apartheid social movements and the quest for the elusive "new" South Africa', *Journal of Law and Society*, 34(1): 77–98.

Markham, A.N. (2013) 'Undermining "data": A critical examination of a core term in scientific inquiry', *First Monday*, 18(10). https://doi.org/10.5210/fm.v18i10.4868

Marston, S.A., Jones III, J.P. and Woodward K. (2005) 'Human geography without scale', *Transactions of the Institute of British Geographers*, 30(4): 416–32.

Massey, R. and Gunter, A. (2020) 'Urban geography in South Africa: An introduction', in R. Massey and A. Gunter (eds) *Urban Geography in South Africa: Perspectives and Theory*, Cham: Springer, pp 1–15.

McCann, E. and Ward, K. (2010) 'Relationality/territoriality: Toward a conceptualization of cities in the world', *Geoforum*, 41(2): 175–84.

McDonald, D.A. (2008) *World City Syndrome: Neoliberalism and Inequality in Cape Town*, London: Routledge.

McFarlane, C. and Silver, J. (2017) 'The poolitical city: "Seeing sanitation" and making the urban political in Cape Town', *Antipode*, 49(1): 125–48.

Miraftab, F. (2020) 'Insurgency and juxtacity in the age of urban divides', *Urban Forum*, 31(3): 433–41.

Moore, A. (2008) 'Rethinking scale as a geographical category: From analysis to practice', *Progress in Human Geography*, 32(2): 203–25.

Mountz, A. and Hyndman, J. (2006) 'Feminist approaches to the global intimate', *Women's Studies Quarterly*, 34(1/2): 446–63.

Ngwenya, N. and Cirolia, L.R. (2021) 'Conflicts between *and* within: The "conflicting rationalities" of informal occupation in South Africa', *Planning Theory & Practice*, 22(5): 691–706.

Odendaal, N. (2019) 'Appropriating "Big Data": Exploring the emancipatory potential of the data strategies of civil society organizations in Cape Town, South Africa', in P. Cardullo, C. di Feliciantonio and R. Kitchin (eds) *The Right to the Smart City*, Bingley: Emerald, pp 165–76.

Parnell, S. and Pieterse, E. (2010) 'The "right to the city": Institutional imperatives of a developmental state', *International Journal of Urban and Regional Research*, 34(1): 146–62.

Pesqueira, L. and Glasbergen, P. (2013) 'Playing the politics of scale: Oxfam's intervention in the Roundtable on Sustainable Palm Oil', *Geoforum*, 45: 296–304.

PoliticsWeb. (2015) 'Cape Town launches Open Data Portal – Patricia de Lille', PoliticsWeb, [online] 27 January, available from: https://www.politicsweb.co.za/politics/cape-town-launches-open-data-portal--patricia-de-l

Pratt, G. and Rosner, V. (2006) 'Introduction: The global and the intimate', *Women's Studies Quarterly*, 34(1/2): 13–24.

Robins, S. (2014a) 'The 2011 Toilet Wars in South Africa: Justice and transition between the exceptional and the everyday after apartheid', *Development and Change*, 45(3): 479–501.

Robins, S. (2014b) 'Poo wars as matter out of place: "Toilets for Africa" in Cape Town', *Anthropology Today*, 30(1): 1–3.

Russell, S. (2013) *The Power of Data as Evidence*, Cape Town: Social Justice Coalition.

Sadowski, J. (2021) 'Who owns the future city? Phases of technological urbanism and shifts in sovereignty', *Urban Studies*, 58(8): 1732–44.

Schäfer, D. (2016) 'Equal Education put safety of schools at risk', PoliticsWeb, [online] 25 October, available from: http://www.politicsweb.co.za/news-and-analysis/equal-education-put-safety-of-schools-at-risk--deb

Smith, J.L. (2022) 'Continuing processes of uneven development in post-apartheid South Africa', *African Geographical Review*, 41(2): 168–88.

Smith, N. (1992) 'Contours of a spatialized politics: Homeless vehicles and the production of geographical scale', *Social Text*, 33: 54–81.

Smith, N. (2004) 'Scale bending and the fate of the national', in E. Sheppard and R.B. McMaster (eds) *Scale and Geographic Inquiry: Nature, Society, and Method*, Malden, MA: Blackwell, pp 192–212.

Smith, R.G. (2003) 'World city actor-networks', *Progress in Human Geography*, 27(1): 25–44.

Social Audit Network. (2016) 'About Social Audit Network', [online], available from: http://socialaudits.org.za/about/

Social Justice Coalition. (2011) 'SJC welcomes significant shift in Cape Town's sanitation policy', [online] 3 October, available from: http://sjc.org.za/posts/sjc-welcomes-significant-shift-in-cape-towns-sanitation-policy

Social Justice Coalition. (2013) 'Report of the Khayelitsha 'Mshengu' toilet social audit', Cape Town: Social Justice Coalition, available from: https://sjc.org.za/wp-content/uploads/2018/11/social_audit_1_sjc_report_of_the_khayelitsha_mshengu_toilet_social_audit.pdf

Social Justice Coalition, Ndifuna Ukwazi and International Budget Partnership. (2015) 'A Guide to Conducting Social Audits in South Africa', Cape Town: SJC, NU, IBP, available from: https://nu.org.za/wp-content/uploads/2021/12/Social-Audit-Guide-e-PDF.pdf

Söderström, O., Blake, E. and Odendaal, N. (2021) 'More-than-local, more-than-mobile: The smart city effect in South Africa', *Geoforum*, 122: 103–17.

South African Cities Network. (2022) 'Who we are', [online], available from: https://www.sacities.net/who-we-are/

Srinivasan, J., Finn, M. and Ames, M. (2017) 'Information determinism: The consequences of the faith in information', *The Information Society*, 33(1): 13–22.

Swyngedouw, E. (2004) 'Scaled geographies: Nature, place, and the politics of scale', in E. Sheppard and R.B. McMaster (eds) *Scale and Geographic Inquiry: Nature, Society, and Method*, Malden, MA: Blackwell, pp 129–53.

van Dijck, J. (2014) 'Datafication, dataism and dataveillance: Big Data between scientific paradigm and ideology', *Surveillance & Society*, 12(2): 197–208.

Watson, V. (2003) 'Conflicting rationalities: Implications for planning theory and ethics', *Planning Theory & Practice*, 4(4): 395–407.

The Worldfolio. (2014) 'Cape Town is one of Africa's most vibrant city destinations', The Worldfolio, [online] 2 October, available from: https://www.theworldfolio.com/interviews/patricia-de-lille-ex/3257/

11

Beyond 'Data Positivism': Civil Society Organizations' Data and Knowledge Tactics in South Africa

Evan Blake, Nancy Odendaal, and Ola Söderström

Introduction

Crises are often defined as unexpected events leading to necessary and urgent action. However, in many cities, especially in the Global South, where large sectors of the population face a series of long-lasting crises in terms of housing provision, jobs, or access to basic services, they also correspond to a situation incorporated in the tactics of everyday life (de Certeau, 1984). South African cities are one of those contexts where crises are not only sudden and unexpected but a constant condition of living and where civil society organizations (CSOs) are intensively involved in their management. A legacy of anti-apartheid struggles, South African cities host a very active network of civil society organizations (Mitchell and Odendaal, 2015; Odendaal, 2021), notably (but not only) in Cape Town: 'it is probably true that Cape Town offers many more pro-poor initiatives and actors than most developing country cities' (Amin and Cirolia, 2018, 283). Since the 2010s, CSOs have used sophisticated data politics to network and establish authority on informal settlement communities. Most recent research on these data politics has focused on Cape Town (Cinnamon, 2020a; Cinnamon, 2020b; Ricker, Cinnamon and Dierwechter, 2020; Pollio, 2022). Cinnamon (Chapter 10, this volume), in particular, has analysed the limits of CSO data-oriented tactics to promote successfully the rights claims of communities towards the state. Our research contributes to this strand of research on South African urban data politics by, on the one hand, extending the study to other South African cities beyond Cape Town, and on the other hand, looking at CSO tactics after the short-lived enthusiasm for data-objectivist tactics in the 2010s.

Our research, in the form of in-depth interviews, participant observation at key meetings, and site work, took place from 2018 to 2020 in South Africa as part of the 'Provincialising Smart Cities in India and South Africa' project, reported on elsewhere (Söderström et al, 2021). We selected three cities as case studies, determined by four criteria. One: a spectrum of cities that can provide many textures of data- and technology-intensive urban policies; two: we desired enough variation between the three city cases to surface the different ways through which such policies are reappropriated or reinvented in varying geographies, which relates to economic profile, settlement typologies, and political leadership. The third and fourth criteria were depth of the empirical material we could gather, and fieldwork access. The three cities under study here – Buffalo City, Ekurhuleni, and Cape Town – complied with these prerequisites. In the second part of this chapter, we discuss the three cases by focusing on the data tactics of one civil society organization (CSO) in each city, emblematic of local data politics, that engages built environment issues, and their relationships to the local state.

In interpreting this work, we are inspired by Warwick Anderson's (2009) reflections on 'conjugated knowledge positions' and by postcolonial science and technology studies (STS). The notion of conjugated knowledge positions, we argue, opens the reflection to data tactics as part of broader knowledge politics and sees them as negotiated within a multi-actor game. In our case, the main actors with whom CSOs negotiate (or conjugate) their knowledge positions are communities they speak to and for, on the one hand, and the local state they speak with, on the other. This analytical lens and the empirical breadth of our study allows us to show that South African CSOs have not rolled out and rolled back data-focused tactics as a consequence of moments of faith and disillusionment in the power of data, but rather mobilize data and other forms of knowledge according to local political contexts and interactional situations. This, we argue in our conclusion, speaks to CSO data politics well beyond the South African case.

On data politics, data activism and the state: a postcolonial STS perspective

While politics around knowledge production and information dissemination is hardly new, the debates surrounding data politics have become increasingly dense and interdisciplinary. Understanding data in context, rather than as a representation of 'facts', is now accepted as part of the ontological and epistemological shifts accompanying datafication: 'Data enacts that which it represents' (Ruppert et al, 2017, 1). This has been enabled through the analytical and algorithmic technologies available to collect and process data, in what Bigo et al (2019) refer to as a broader historical transformation where the exchange of data potentially reconfigures the relationships between actors implicated in its collection, production, and representation. Not only can data not be disassociated from surrounding social and political struggles, but to attempt to do so would be to mask such contestations (Ruppert et al, 2017).

Access to the means to produce data, and increasingly individualized control of its representation and dissemination, has, within the context of the theme of this chapter, resulted in data as a currency in the relationship between the state and civil society (Beraldo and Milan, 2019, 5). It is now generally accepted that data has performative power that has political consequences. By contributing to redefine how citizenship can be enacted in the age of datafication, our study provides a concrete example of how data politics from the ground up potentially 'reconfigures relationships between states and citizens'. The availability of data, the shifts from demographic approaches to individualized targeting, and the opacity and power of computational modelling has led to a terrain of new players and power brokers in data-driven state affairs.

The ubiquity of tools available for knowledge co-production has created opportunities for a shift from data politics to the politics of data, or what Beraldo and Milan refer to as the 'contentious politics of data' (2019). The appropriation of data and associated practices outside the state and corporate sector has the potential to impact power relations. In our work we are particularly interested in the tensions between city discourses and knowledge generation by CSOs. This becomes particularly poignant when considering the collective power of data and considerations of agency. The availability of data makes many iterations of agency possible through social practices, rooted in technology and social networks. Thus, context is important, as are data cultures.

It is worth dwelling on the genesis of work on data and mobilization since it does have a bearing on these data cultures. The area of data politics is by its nature interdisciplinary. In media studies, we have long encountered the power of citizen media and empowerment strategies, with long traditions of alternative, independent, and community media that bring citizens to the forefront of knowledge production. This arena, where journalism meets data, raises interesting opportunities for understanding how CSO data practices interface with traditional media and build on such histories. Histories of activism shape data cultures, which in turn shape the protocols, framings, and subject definitions that inform the collection and dissemination of information. Such data practices are associated with social practices and agents (Ruppert et al, 2017). Data regimes can be arenas of contestation, as interest groups may disagree on social categories and definitions, or certain data practices may be restricted (Bates, 2018). Ethical and social concerns are important underpinnings of processes of production, processing, and distribution of information. Bates refers to the combination of such ethics and the range of coding and classification norms and practices, as 'data cultures' (2018, 191) that are deeply contextualized through cultural norms, value systems, and beliefs.

Recent work on data power and politics goes beyond Western/Northern analyses. Historically, the design of statal data regimes have been relationally developed between metropoles and colonies, as mapping and censuses were often designed and first deployed in colonial territories before being used in the metropole. Today, data politics cannot be envisaged within national or regional containers but as relational: 'Just as a quest for an imperial census happened simultaneously to the development of national modern censuses, so too are quests to know populations through big data happening transnationally' (Isin and Ruppert, 2019, 219). Past and present data politics in former colonial territories are thus entangled with data collection and management processes crafted in former metropoles. They are also shaped by persistent but reconfigured asymmetries of power.

If data politics in the Global South must be situated in a global context, they must also be envisaged in their specificity. Milan and Treré (2019, 319) have thus called for a 'de-Westernization of critical data studies' to account for how datafication unfolds in the Global South. This involves understanding the Global South as a 'composite plural entity', 'bringing agency to the core of our analyses' and 'zooming in on the specific ways of thinking data from the margins' (324). The uses of data are relational in many ways: they are shaped by (post)colonial logics, by political and historical local contexts, and the strategies and tactics of CSOs.

Recent work on social audits in South Africa has shown that municipal open data policies have played an important role in data politics around informal settlements (Ricker et al, 2020). 'Seeing like the state' is strategically used by CSOs to persuade the state (McFarlane and Silver, 2017), particularly in a context where the South African government is mandated by the Bill of Rights in the Constitution to provide access to shelter and basic services. Cinnamon (2020a) argues that this reveals a belief held by CSOs in data and quantitative evidence as central elements in policy making and grassroots activism, which tends to reproduce governmental smart cities discourses. The scalar tactics of grassroots organizations, whereby local hardships would be made visible through data to a 'more-than-local' public, did not have the desired impact (Cinnamon, Chapter 10). This has led to growing scepticism amongst CSOs with regard to the power of data-focused strategies. Data is not enough to influence decisive changes in informal settlements. A similar scepticism has developed regarding how social audits can amend and improve municipal data through the inclusion of CSO and community-collected data. A study of three CSOs in Cape Town indicates that while municipal data is available, there is no true bidirectional open data policy (Ricker et al, 2020).

What we find of considerable interest, given this critique, is how data tactics are combined with more traditional practices in the collection, packaging, and representation of information. Following Ruppert et al (2017, 1) we are not only interested in how data enacts what it represents but the influence and framing through everyday practices of resistance and mobilization. How can this framing impact public policy and substantive goals such as spatial justice? We respond to the call for more in-depth research in this area (Ricker et al, 2020, 371), notably by paying attention to 'activism's chameleonic quality' (Cinnamon, 2020a, 636). Through its reactivation and redefinition of citizenship, data activism engages the lived experiences of individuals and does so in the framework of collective experience (Beraldo and Milan, 2019). Milan and Gutiérrez (2015) refer to the notion of 'connective action' that moves beyond the atomistic-community tension, as a means through which 'lived experiences' are enrolled in data practices. Understanding the representation of these in interacting with the state requires empirical investigation and the mobilization of appropriate theoretical resources.

In an exploration of the relationships between STS and social movement theory, Hess (2016, 518) focuses on technology as comprising 'material objects that are intentionally used to modify the social and/or material world', rather than be the subject of focus or change. Within this frame, for technology to be effective as having mobilization agency, it is best understood as embedded in socially and historically situated cultural practices (Monahan, 2005), as part of a sociotechnical system (Hughes, 1987), a web of human–object relations (Bijker and Law, 1994), or a network of persons, institutions, and things (Callon, 1984). In understanding the materiality of datafication and associations with the meaning-making practices of local struggles, STS adds a sensibility towards understanding material-semiotic processes, with no predefined objects of analysis. Politics can manifest in the design of technology design, related policy mobilization, and in patterns of use. The STS study of the entanglement of technology and politics has long been rather insensitive to postcolonial or decolonial perspectives (Söderström, 2018).

However, Law and Lin (2017) have recently argued for a postcolonial version of the STS term 'symmetry', a conceptual space of multiple centers, many points of disjuncture, and no single postcoloniality. Anderson (2009; 2017) refers to his own work using 'Asia as method' as a means to decenter STS. We would argue that many frames of reference and many post-colonies rub up against one another, exposing the limitations of thinking of South Africa as one Global South entity. A critical engagement with the heterogeneity within acknowledges internal contradictory epistemologies and interpretations of sociotechnical solutions to urban problems. A postcolonial analysis offers a 'flexible and contingent framework for understanding contact zones of all sorts, for tracking unequal and messy translations and transactions that take place between different cultures and social positions' (Anderson, 2009, 395).

In our case, these contact zones relate to triangular relations between the state, CSOs, and community. In these relations, how each party considers the others, and the validity and legitimacy of their knowledge plays a crucial role in determining the efficacy of CSO initiatives. This asymmetry was theorized famously by Foucault (1980) as a power-knowledge question in which some, like the mentally ill, occupy a 'subjugated knowledge position'. The task of the analyst was, for Foucault, to make forms of subjugated knowledge visible and 'capable of opposition and struggle against the coercion of a theoretical, unitary, formal and scientific discourse' (1980, 85). However, imported into the terrains we are studying, this notion of subjugated knowledge could lead to an essentialized understanding of the knowledge produced by CSOs and communities. There is no such thing as a pure grassroots knowledge of the community independent of interactions with other actors and knowledges. This risk of essentialization has been recognized more generally when decolonial thinking does not inscribe 'indigenous' knowledge in circuits of globalization (Streule et al, 2020). Therefore, scholars in postcolonial STS have argued that knowledge and knowledge production should always be understood as the product of entangled histories, making it necessary to consider the subjects of knowledge as *conjugated*, rather than simply *subjugated* (Anderson, 2009). Our position is that data politics can only be understood in relation to context-dependent conjugated knowledge positions. Our cases, discussed below, study these relations comparatively.

South African ICT and data–driven governance context

The current policy frame for the development of data–driven governance in South Africa refers to a number of ICT and data-focused themes that permeate national discourse. The first relates to sector-specific strategies as they relate to infrastructure rollout and economic targeting and innovation. Examples include the need for smart infrastructure to enable human development and as an important economic driver in the National Development Plan (RSA, 2013), in some cases referred to as 'economic infrastructure' (pp 63–64). Governance is an important theme that emerges in the Integrated Urban Development Framework (RSA, 2016), focusing on ICT as a connector and integrator across infrastructure platforms. Sector-specific strategies such as the ICT Roadmap, the National Transport Master Plan and the South African Smart Grid Initiative see ICT as catalytic to innovation and service efficiency.

Up until 2018, the smart city idea has been largely implicit in national policies and plans. However, in a review of speeches and political media claims, the notion of the Fourth Industrial Revolution emerged more frequently. This culminated in an official stance communicated in 2019 with President Cyril Ramaphosa encouraging the construction of smart cities as new towns. The commitment to the construction of smart cities was further elaborated on in the 2020 State of the Nation Address. Given the unemployment rates and economic difficulties experienced, the job creation potential of new technologies and associated industries is articulated in economic as well as human development terms.

Any implementation of such policies would need to be driven by local government, however. The South African constitution provides for a number of measures that have implications for new urban extensions and investments. Municipalities are autonomous and are largely responsible for generating their own budgets. This decentralized system is political (with separate local government elections), financial, and ideological. Local government is mandated to be developmental as the state apparatus 'closest to the people' and responsible for service delivery as well as local economic development (RSA, 1996). Furthermore, municipal legislation includes stringent procedures for enabling a participatory democracy and collaborative plan-making (RSA, 2000). In addition to that, convoluted procurement procedures and stringent anti-corruption measures constrain top-down and/or corporate-led urban development and would need to be budgeted for and agreed to in cities' integrated development planning processes. The integrated development plan (IDP) is essentially a municipality's business plan and is reviewed every five years. Preparation of this plan is required by law, and is required to be participatory and representative of a long-term development vision. Thus, each of the three cities discussed here – Buffalo City, Cape Town, and Ekurhuleni – would be impacted by the municipal strategic goals. Conversely, each is also impacted by how well or how badly it performs against these goals. This is often where civic action becomes critical to service delivery, as civil society organizations (CSOs) use data strategies to confront local government.

Here, we compare the data perspectives of three CSOs correlating to the governance contexts of Cape Town, the Gauteng City-Region, and Buffalo City. These organizations differ in their histories, their principal objectives, and operational methodologies. Despite differences, they share two characteristics: a focus on the built environment, and some form of engagement with the local authority on data. Their tactics are also emblematic of CSO tactics in each city.

City of Cape Town: a metro using data approaches to manage climate crisis

The City of Cape Town, located in South Africa's Western Cape province with a population of approximately 4.8 million people, is the second largest metro and parliamentary capital city in South Africa.

Politically controlled by the national opposition, the Democratic Alliance (DA), since 2006, the city has predicated its governance identity on mission statements of 'the city that works for you' (de Lille, 2014) and, more recently, a city that is 'accountable and responsive to its residents' (City of Cape

Town, 2021). Such monikers serve to position the DA in contrast to the alleged corruption, misspending, and governance mismanagement under the National Government (Democratic Alliance, 2013, 7). But beyond political soapboxing, these slogans also articulate a governance ambition guided by data-fuelled empiricism and evidence-based decision-making, particularly in the face of crisis.

The City has an established history of infusing its spatial plans with evidence-based decision-making approaches (Odendaal and McCann, 2016). Led by the former mayor, Patricia de Lille, the City pursued an open data policy with the intent of guiding public, academic, and free market solution ideation through access to government data (City of Cape Town, 2016). This policy served as a double-edged sword for the City, equipping CSOs with municipal data to supplement their own community-sourced intelligence on service delivery mismanagement (Ricker et al, 2020). More recently, the management of a 'day zero' drought crisis during 2018 brought political attention to the necessity of a strategic, data-driven overlay to address cross-cutting governance issues. Sitting above and cutting across the City structures, a Resilience Strategy fed multiple points of quantitative data and grassroots barometers into a series of dashboards. This management of the drought crisis during 2018 foregrounded the role of data to the City's leaders, generating political will for the development and public launch of a Data Strategy in 2020 (City of Cape Town, 2018).

Serving as an evolutionary tributary to the Resilience Strategy, the Data Strategy focuses on internal, strategic affairs within the City. Focused primarily on internal management matters, the strategy aims to mediate a centralized, coordinated set of data relations across City departments. It conceptualizes data as a collective resource: collating multiple streams of data with an aim to improve standardization in data management across City operations. With this internal focus, the strategy does not make explicit provisions to acknowledge and engage with grassroots-sourced data from community activists and advocates. The strategy does, however, make some provisions for CSO partners to interface with the data strategy as a source of supplemental evidence-based intelligence. Below, we show how tactics of CSOs in Cape Town engage with this municipal strategy to affirm its position as legitimate knowledge provider to promote rights of the community.

Violence Prevention through Urban Upgrading (VPUU): Data-driven mediation between local government and communities

Violence Prevention through Urban Upgrading (VPUU) is a non-profit organization established in 2005 in Cape Town. The organization promotes the design of public environments and associated governance arrangements to enable greater safety in former townships in the south-east of the city. VPUU's spatial planning methodology has since inception used an evidence-based decision-making approach (Brown-Luthango et al, 2017; Ewing and Krause, 2020; Ewing, 2021). More recently, it adopted an ICT for Development (ICT4D) approach: first, to advocate for ICT skill and infrastructure development in partner communities; and second, to develop protocols and metrics that enable ongoing monitoring of the efficacy of development interventions through community-drawn data. ICT4D principles serve as a core component to VPUU's organizational approach, helping to define data-driven 'intelligence' as a recursive process for both CSO and the state:

'We want to be seen building a capacity and building a legacy in the community. And that for me requires us to answer, "How can you track that legacy?" You need to have data. You can call it evidence; you can call it all kinds of things [...] I'm not interested in policy. For me it's about implementation and learning the hard way. And implementation is not just about building a facility, it's about pulling the City accountable and saying to them, "You have the resources to operate and perform maintenance in the long term". That is also what seeing the outcome is about, to see if it was a positive or a negative.'
(Interview with a senior member of VPUU, 2019)

VPUU has developed a data infrastructure to support its aims. VNET serves as a data science ecosystem to facilitate the gathering, synthesis, analysis, and output of data as community 'intelligence'. It is composed of nodes established around a community site, primarily in spaces where VPUU hosts interventions.

Functioning as a mesh network, these nodes serve to wirelessly direct collected data via app-based interfaces back to a central hub within a community. VPUU assists in the construction of community resource centres, serving as public spaces for hosting skills development programmes as well as providing a computer centre and access to public Wi-Fi (Figure 11.1). Community centres also serve as central hubs in the network, transmitting data back to VPUU's central offices via fibre-optic line. A team of data scientists and GIS trained technicians work with this transmitted data, generating 'intelligence'. In particular, VPUU's Community Atlas and WaziMap data platforms combine community-sourced data with supplementary and official data sets in order to establish a longitudinal approach that diligently records and archives data and informs decision-making systems (Pillay, 2019).

This data-driven approach extended with how the organization combines an explicit community development agenda with VNET. The rollout of technical equipment is matched with workshop interventions, serving to

Figure 11.1: VPUU members at the community centre in Monwabisi
Park, Khayelitsha

Source: Ola Söderström.

instruct community members of the purpose of network infrastructure.
Community members are trained by VPUU as data collectors, using bespoke
application software on mobile devices to capture specific metrics of service
delivery. The metrics for this collection and measurement are decided on
in consultation with community members, but primarily informed by best
practice in the field of intervention. However, this subtle undercurrent of
'community input as support to technical expertise' surfaces an uneven
politics that reinforces the power allocated to abstracted data.

For VPUU, the focus on 'intelligence' and evidence-based decision-
making is intended to foreground accountability, serve as a device for
monitoring and evaluation, and increase the legitimacy of the CSO. Data
thus enables trust between the state and community. VPUU uses its data
infrastructure to illustrate what development models are best suited to making
actionable, practical steps towards 'safe, sustainable, integrated communities,
citizenship, pride and the improvement of quality of life for residents' (VPUU,
2021). Cast within the organization's VNET, data serves several important
functions: collaboratively identifying key issues, formulating solutions, and
co-developing leadership structures to engage with the state.

The empiricism and claims to neutrality that guide the technical expertise
of staff manifests as a decidedly objectivist undercurrent. Community
knowledge is given currency for state mediations when transmuted through
the lens of technical expertise and data objectivity. Community knowledge

becomes supplementary to social development rather than the essence of the advocacy itself. Technical, quantitative, evidence-generating systems are the primary means to leverage the state. Community members are, however, not passive in this dynamic. They are recognized by CSOs as partners (along with the state) in shaping the direction of development interventions, highlighting gaps, and directly benefiting from their success. Thus, VPUU's tactic speaks the language of 'dataism' (van Dijck, 2014) or 'data positivism' (Power, 2022). It is a conjugated knowledge position where knowledge about the community mainly takes the form of data points collected and analysed by the CSO to speak to the local state in its preferred language. Although such data-objectivism can always be under attack when practiced by a CSO and has been less successful than expected (Cinnamon, 2020a, Cinnamon, Chapter 10), CSOs like VPUU have kept their faith in the efficacy of this tactic in the specific context of Cape Town. However, our two other case studies show that other tactics are deployed in other contexts in South Africa today.

Gauteng City-Region: a conurbation focused on catalytic economic development

The Gauteng City-Region (GCR) refers to the cluster of cities and towns forming the economic powerhouse in the interior of South Africa. Notable urban cores include Tshwane, Johannesburg, and Ekurhuleni. Together, this conurbation accounts for almost 16 million habitants (Gauteng City-Region Observatory, 2022). Although administered as different municipal centres, there is recognition that the large-scale economic activities of any one urban centre have implications for the collective region and the country as a whole. Across the last twenty years, ICT developmentalism has shaped governance of this region. In particular, ICT infrastructure development, e-governance projects, and catalytic development agendas have shaped the agenda of a region looking to cement its economic status through smart strategies and ICT-driven development (Söderström et al, 2021).

In Johannesburg, the last 15 years has seen efforts to develop ICT access in underserviced areas. Such a network aimed to establish Johannesburg's image as a world-class city with established ICT infrastructure capable of stimulating the economy (Ericsson, 2009). During the same time, the city of Ekurhuleni municipal government made use of e-governance to focus on developing internal ICT capacities in order to create a digital front to expedite service delivery to constituents. It was primarily a public-focused operation concerned with the mundane logistics of governance. The governance logic underpinning this deployment of data-tracking is to maintain a standard of service delivery for more economically privileged residents to ensure a steady income stream that can, in turn, be directed towards underserviced communities (Mail & Guardian, 2011). Ekurhuleni, which hosts the country's main passenger and freight airport, has also been the site of catalytical development delivered via an 'aerotropolis' project. The aerotropolis is positioned as a state-driven initiative that spatializes distinct nodes for economic development, including 'aviation-intensive transport, telecommunication, accommodation, commercial, logistics, industry and related enterprises [...] entail[ing] investment on new economic infrastructure to support logistics, distribution and related green industries' as Mondli Gungubele, Executive Mayor of Ekurhuleni, explained in 2011 (Moeng, 2011). Utilizing the position of the airport and downstream industries, the aerotropolis establishes a logic through which ICT-based airport-focused strategies can generate economic growth.

In the face of resource and capacity constraints, the concept of the e-governance and the aerotropolis offers an alluring neoliberal remedy for beleaguered local governments to address regional social and economic development via 'redirected' downstream benefits: a priority placed above face-to-face engagement over key governance issues and policy agendas with an historically disenfranchised urbanite majority.

The CSO Planact has developed its knowledge tactic in this context, playing the data game quite differently compared to VPUU, as we show below.

Planact: Building a foundation for participatory governance via grassroots knowledge politics

Planact is a built environment non-governmental organization that emerged in Johannesburg during the later apartheid years. Planact has a reputation for acting in strong advocacy for the interests of the marginalized, having provided technical and planning support to poor communities since the 1980s. The organization's approach to data displays a commitment to community self-representation and self-identification through community mapping. This data practice is reinforced by sustaining interpersonal and cordial relations with local government actors in the Gauteng City-Region and the politicization of community knowledge. The organization has found renewed purpose in the post-apartheid era as a community development mediator. It retains the professional sensibility rooted in its founding years but integrates its technical know-how with community-centred advocacy and a broader civil society agenda (Planact, 2021):

> 'Participatory governance, I think, is where the crux of the work that Planact does. This is where we work with social movements and community-based organisations – just for them to understand how governance works from the inside.'
>
> (Interview with Planact senior member, 2019)

Rather than rely on demonstrable evidence to engage the state (whether it be of service delivery failure or a successful implementation of an intervention model), Planact staff network with state actors directly, forming one-on-one relations with government officials and politicians. The interpersonal relationships with elected officials create spaces to elevate the value of the experiences, insights, and capabilities of communities. Thus, knowledge practices at the grassroots are made visible and legitimized in relation to 'formal' governance.

Community mapping and street-naming interventions initiated by Planact in Skoonplaas settlement, located in Ekurhuleni, are examples of community-centred knowledge politics. These interventions serve as an alternative and supplementary exercise to traditional enumeration practices. Household enumeration is used by NGOs and local government structures to record and itemize residential typologies and forms of tenure, typically in informal settlements otherwise rendered invisible by official maps (Baptist and Bolnick, 2012). Planact's approach uses collaborative mapping as a means to make the invisible visible. In this context, community mapping is largely a physical, pen-and-paper affair (Figure 11.2). Community members collectively identify themselves through physically inscribing new street names, routings, neighbourhood spaces, and boundaries onto hand-drawn maps. These processes require community members to engage with one another on how best to define territory, which capacities and histories are

Figure 11.2: Community training to canvass the settlement of Skoonplaas for a street-naming intervention

Source: Planact, 2019.

mapped, represented, and made visible, and how best to demonstrate this to others.

Planact's explicit aim in developing such forms of self-representation is for community members to negotiate forms of self-recognition first among themselves, and in relation to the outside. In effect, this mapping builds a grassroots foundation for participatory governance. There is of course a functional dimension also: the legibility enabled through street-mapping facilitates emergency response services. Ascribed street names and numbers equip service providers with the insider knowledge necessary to navigate otherwise labyrinthian networks of streets and pathways. Thus, this mapping data is not disseminated to the state as a piece of open data. It is given out by the CSO with a specific political intent: to call on government support, when necessary, with very specific information that assists the community on their direction.

The organization avoids taking an adversarial approach to its negotiations with the state. It favours interpersonal, culturally sensitive relationships with officials:

'We are learning the roots on how mediating these institutions happens. Like, you know, you can use confrontation, but you also have to be smart. I have to use Zulu now, you know. I think what we are learning is that language is important. Culture plays an important role, you know, and perceptions and all those negative things that we normally call soft skills or soft issues. They all come into the play. When you engage with an official, you need to be mindful of all this because, you know, if you are not, he shuts the door in your face. He [an official] listened, you know. But it took us two or three good meetings for him to really absorb what we are talking about. And even more effort to even secure a meeting.'

(Interview with Planact senior official, 2019)

Equipped with Planact's institutional knowledge of municipal processes and a sense of the general mood of officials, the CSO and community representatives closely collaborate on negotiation tactics and selecting intervention results to bring to the table.

The claims of visibility through local experience – and the political space it teases open for community knowledge in urban development agendas – has implications for the relationship between CSO, state, and community. These are deeply political processes through which conjugated knowledge positions are developed. Community-based resources, networks, and partnerships acknowledge community knowledge as a means in of itself to address locally defined issues. If data as evidence is central to VPUU's knowledge politics, shared knowledge at community level is central to

Planact's tactics in Ekurhuleni. The third case we investigate occupies an intermediate position between these two poles, further extending modes of conjugated knowledge.

Buffalo City Municipality: a metro on the precipice of governance crisis

The Buffalo City Municipality (BCM) is located along the coastline of South Africa's Eastern Cape province. With a population of approximately 700,000, this smaller metro is often overlooked in urban development discourse compared to larger South African cities (Söderström et al, 2021). A conurbation of towns and cities such as East London and Bhisho, BCM and its surrounds have long been regarded as the country's automotive industry heartland and an important political homeland for the ruling African National Congress. The expansion of the metropolitan area and incorporation of the surrounding rural fringe to become Buffalo City has historical context: addressing racialized unequal development between the industrialized core of East London at the coast and the former homeland reserves in the hinterlands.

Considering this context, BCM's development agenda is focused on housing and service delivery and job creation (Sikhakane and Reddy, 2009). To spur on job creation, the town of East London hosts one of South Africa's Industrial Development Zones initiated by the national government to bring foreign direct investment and develop export-orientated industrial development within targeted regions.

While such projects focus on regional-scale economic development goals, service delivery agendas at the local scale have been constricted by persistent municipal and provincial mismanagement and corruption (Basopu, 2016), mired by a history of underfunded municipal operations in the Eastern Cape (Sikhakane and Reddy, 2009), and a dysfunctional devolution of power across scales of local government (Masango et al, 2013). These constrained operational contexts and a perpetual crisis of municipal mismanagement shape how data is defined within government. Municipality-driven ICT and data initiatives are ostensibly focused on smart infrastructure upgrades and internal municipal connectivity. However, the hierarchical and siloed nature of BCM management has created a governance environment not receptive to data- and knowledge-sharing (Ncoyini and Cilliers, 2020). Municipal departments thus practice competitive, siloed protectionism to guard custom-tailored systems and prevent disruption to their baseline operations. These conditions have generated disconnected relations between line departments and with the political municipal core. It is in this context of economic crisis and political fragmentation that Afesis-corplan has developed its data and knowledge tactics.

Afesis-corplan: Technical mediation to support capacity-building in under-resourced contexts

Afesis-corplan, active in the Eastern Cape since the 1980s, focuses on enabling a participatory democracy comprising communities and local government, informed by a 'theory of change' that balances technical and institutional knowledge with local experience (Afesis-corplan, 2021a). This change is enabled through incrementalism: multimodal, step-by-step approaches that focus on communities as the substrata of institutions that enable social development from the bottom up. This is reflective of a broader trend in South African data politics. Understanding the operational stresses that face municipal governments, CSOs engage with the state to offer a blueprint for step-by-step, collaborative community development. Data has a particular role in this blueprint, providing an empirical justification for the claim-making and community advocacy put forward by the CSO.

Afesis-corplan uses data as a nexus point to draw disconnected governance actors (community, CSO, and state) into relation: formulating community insight parsed through CSO technical knowledge to deliver practical models that address persistent slow service delivery from local government. In delivering on this mandate, sustaining formalized, routine relationships with municipal line departments serves to engender a responsive and receptive engagement with the state:

'We're not necessarily antagonistic towards local government. If anything, I think we have positioned ourselves by virtue of what we do. We find ourselves being in that space whereby we're conscious of the concerns of the community and also with the local municipality. When we sit down with the [Buffalo City] municipality, you get to understand their thinking and their opinion of their projects and why certain things are not happening the way that they're supposed to. And in that, you are able to do much more for the community, because you're able to give them detailed information and plans influenced by knowing what is happening inside local government.'
(Interview with Afesis-corplan senior member, 2019)

Afesis-corplan surfaces community knowledges through trust-building with communities and then, combines these with the organization's own technical expertise in settlement planning. Staff first embed themselves in the community structures of informal settlements to develop relationships of trust. Livelihood histories and needs assessments then follow, collected directly from communities and used to apply pressure on elected officials to address service delivery. More strategically, though, collected insights

Figure 11.3: A path towards a partnership-driven, incremental housing delivery model

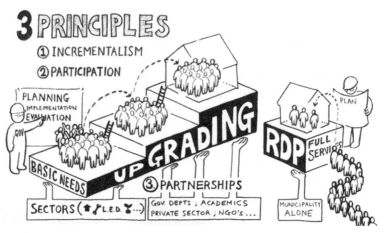

Source: Afesis–corplan, 2016.

(both qualitative and quantitative in nature) directly inform an evolving plan within the organization's technical base for an incremental, 'stepping stone' housing model (Figure 11.3). This model aims to move people from a state of being tenure insecure towards a consolidated state of owning title deeds for property with linked basic services (Afesis–corplan, 2021b).

This incrementalism is predicated on a concrete and embodied understanding that:

> 'It's not just focused on the house as a product, but it's focused on housing as a process. It's to say, "How do you look at the person's life who is staying there post the project?" You don't just give them a house and then you live there, but you're equipping them with something, you're giving them a skill that they can use at any stage. It's a spectrum.'
>
> (Interview with Afesis–corplan senior member, 2019)

Thus, a blueprint to multiple approaches to housing delivery across a spectrum of needs is an enticing hook to capacity-scarce government departments, demonstrating that not everyone requires the same degree or form of government support at the same time:

> 'It reduces the responsibility on the government's side as well and the expectation placed on government. It creates a situation where people can say "You've given me recognition of tenure, here's my contribution towards this."'
>
> (Interview with Afesis–corplan senior member, 2019)

Afesis-corplan uses its data collection and knowledge-building strategy to shape the terms of engagement with BCM over development initiatives like incremental housing. Afesis-corplan acknowledges that the training and background of staff members as professional planners provides some technical legitimacy with local government officials, overcoming accusations of subjective claim-making. This shared technical foundation between state and CSO thus confers community knowledge with a greater degree of weight. This positions Afesis-corplan as a legitimized source of community information and development insight to which municipal planners and line department staff do not have access. In the language of actor–network theory, Afesis-corplan acts as a mediator rather than a simple intermediary (Latour, 1987): it modifies community knowledge to make it amenable to the technical language of planning. While VPUU speaks the language of the state and Planact gives priority to the language of the community, Afesis-corplan's conjugated knowledge position focuses on a translational exercise.

Conclusion

The three South African case studies discussed represent a range of historical, political, and geographic milieus that impact the data politics of local CSOs. They also represent an opportunity to examine data tactics as contextually driven, informed by differing perspectives. These insights into how we think about data are helpful ways to conceptualize, as our empirical work shows, the practices of CSOs engaged in urban development issues in South Africa: VPUU's data-driven approach to development mediation offers a response to a governance context defined by explicit valorization of evidence-based decision-making.

In the Gauteng City-Region, Planact responds to a governance environment that has become too bloated, detached, and focused on large-scale developmentalist trajectories at the risk of losing sight of the core community tenants of participatory democracy at its most basic level. Rather than engage the local state head-on with data games, Planact instead emphasizes grassroots knowledge politics to resurface the potential for human-scaled democratic practices between government and community. Afesis-corplan produces data-supported, capacity-generating development models to support the work of a failing local state where the focus for economic salvation lies in a regional economic catalyst that is detached from the everyday realities of basic service provision and housing. Our case studies reveal a form of localization of data politics from within: uneven and variegated data perspectives deployed to interface with different governance contexts. Each tactic is the result of a context-specific triangulation between CSOs, communities, and the local state.

The notion of conjugated knowledge position has been helpful in our comparative study to identify rather subtle differences between South African urban data and knowledge tactics. These subtle differences reveal that beyond 'data universalism' (Milan and Treré, 2019), even in one specific context such as South Africa, there is no such thing as a single 'Southern urban data politics'. The mobilization of data, as standardized and abstracted information, is, in the present post social audit phase in South Africa, variegated. Urban data and grounded knowledge are mobilized alternatively, and more or less translated, as responses to municipal context and depending on the aims of interactions, from community-building to housing delivery planning. These tactics, determined by a pragmatism of what works in situations of long-lasting crises, should invite us not only to attend to the variety of Southern urban data politics but also to the limits of reducing contemporary urban data politics to a generalization of surveillance infrastructures and the triumph of 'data positivism' (Power, 2022).

Acknowledgements

We would like to acknowledge the generous support of the Swiss National Science Foundation (SNSF). This chapter is one of the results of the SNSF-funded research project Smart Cities: 'Provincializing' the global urban age in India and South Africa (grant 10001AM_173332).

References

Afesis-corplan. (2016) 'A policy and strategy for upgrading informal settlements', [online video] 2 February, available from: https://afesis.org.za/programmes/sustainable-settlements/

Afesis-corplan. (2021a) 'Company profile', [online], available from: https://afesis.org.za/wp-content/uploads/2019/01/Company-Profile-2016_web.pdf

Afesis-corplan. (2021b) 'Upgrading informal settlements, [online], available from: https://afesis.org.za/upgrading-of-informal-settlements/

Amin, A. and Cirolia, L.R. (2018) 'Politics/matter: Governing Cape Town's informal settlements', *Urban Studies*, 55(2): 274–95.

Anderson, W. (2009) 'From subjugated knowledge to conjugated subjects: Science and globalisation, or postcolonial studies of science?' *Postcolonial Studies*, 12(4): 389–400.

Anderson, W. (2017) 'Postcolonial specters of STS', *East Asian Science, Technology and Society: An International Journal*, 11(2): 229–33.

Baptist, C. and Bolnick, J. (2012) 'Participatory enumerations, in situ upgrading and mega events: The 2009 survey in Joe Slovo, Cape Town', *Environment & Urbanization*, 24(1): 59–66.

Basopu, P.M. (2016) 'Critical assessment of corruption in municipalities and its impact in service delivery: Case study Buffalo City Metropolitan Municipality', PhD thesis, Alice, SA: University of Fort Hare.

Bates, J. (2018) 'Data cultures, power and the city', in R. Kitchin, T.P. Lauriault and G. McArdle (eds) *Data and the City*, Abingdon: Routledge, pp 189–200.

Beraldo, D. and Milan, S. (2019). 'From data politics to the contentious politics of data', *Big Data & Society*, 6(2). https://doi.org/10.1177/20539 51719885967

Bigo, D., Isin, E. and Ruppert, E. (2019) 'Data politics: Introduction', in D. Bigo, E. Isin and E. Ruppert (eds) *Data Politics: Worlds, Subjects, Rights*, London: Routledge, pp 1–19.

Bijker, W.E. and Law, J. (eds) (1994) *Shaping Technology/Building Society: Studies in Sociotechnical Change*, Cambridge, MA: MIT Press.

Brown-Luthango, M., Reyes, E. and Gubevu, M. (2017) 'Informal settlement upgrading and safety: Experiences from Cape Town, South Africa', *Journal of Housing and the Built Environment*, 32(3): 471–93.

Callon, M. (1984) 'Some elements of a sociology of translation: Domestication of the scallops and the fishermen of St Brieuc Bay', *The Sociological Review*, 32(1 suppl): 196–233.

Cinnamon, J. (2020a) 'Attack the data: Agency, power, and technopolitics in South African data activism', *Annals of the American Association of Geographers*, 110(3): 623–39.

Cinnamon, J. (2020b) 'Power in numbers/Power and numbers: Gentle data activism as strategic collaboration', *Area*. https://doi.org/10.1111/area.12622

City of Cape Town. (2016) 'City of Cape Town Open Data Policy (Policy Number 27781)', [online] 26 May (revised 3 December 2020), available from: https://resource.capetown.gov.za/documentcentre/Documents/Byl aws%20and%20policies/Open_Data_Policy.pdf

City of Cape Town. (2018) City of Cape Town Data Strategy, available from: www.capetown.gov.za/councilonline_layouts/OpenDocument/Open Document.aspx?DocumentId=42cfdac5-a6e3-498d-942c-cb3d80ccf80d

City of Cape Town. (2021) 'Council Speech by the Executive Mayor, Alderman Dan Plato', [online] 25 May, available from: https://www.capet own.gov.za/Media-and-news/Council%20Speech%20by%20the%20Ex ecutive%20Mayor,%20Alderman%20Dan%20Plato

Democratic Alliance. (2013) 'DA Policy on Governance', available from: https://cdn.da.org.za/wp-content/uploads/2018/02/14234256/Gove rnance1.pdf

de Certeau, M. (1984) *The Practice of Everyday Life*, translated by S. Rendall, Los Angeles: The University of California Press.

de Lille, P. (2014) 'Cape Town's new logo and slogan explained', PoliticsWeb, [online] 24 February, available from: https://www.politicsweb.co.za/polit ics/cape-towns-new-logo-and-slogan-explained--patricia

Ericsson. (2009) 'Joburg kicks off digital project', ITWeb, [online] 24 April, available from: https://www.itweb.co.za/content/VgZeyqJVkdNqdjX9

Ewing, K. (2021) 'Spaces of transformative practice: Co-producing, (re) making and translating fractional urban space in Gugulethu, Cape Town', *Urban Forum*, 32: 395–413.

Ewing, K. and Krause, M. (2020) 'Emthonjeni – Public space as smart learning networks: A case study of the violence prevention through urban upgrading methodology in Cape Town', in A. Aurigi and N. Odendaal (eds) *Shaping Smart for Better Cities: Rethinking and Shaping Relationships between Urban Space and Digital Technologies*, Amsterdam: Elsevier, pp 339–56.

Foucault, M. (1980) 'Two lectures', in C. Gordon (ed) *Power/Knowledge: Selected Interviews and Other Writings 1972–1977*, New York: Pantheon Books.

Gauteng City-Region Observatory. (2022) The Gauteng City-Region, [online], available from: https://www.gcro.ac.za/

Hess, D.J. (2016) *Undone Science: Social Movements, Mobilized Publics, and Industrial Transitions*, Cambridge, MA: MIT Press.

Hughes, T.P. (1987) 'The evolution of large technological systems', in W.E. Bijker, T.P. Hughes and T. Pinch (eds) *The Social Construction of Technological Systems*, Cambridge, MA: MIT Press, pp 51–82.

Isin, E. and Ruppert, E. (2019) 'Data's empire: Postcolonial data politics', in D. Bigo, E. Isin and E. Ruppert (eds) *Data Politics: Worlds, Subjects, Rights*, London: Routledge, pp 207–27.

Law, J. and Lin, W. (2017) 'Provincializing STS: Postcoloniality, symmetry, and method', *East Asian Science, Technology and Society: An International Journal*, 11(2): 211–27.

Latour, B. (1987) *Science in Action: How to Follow Scientists and Engineers through Society*, Cambridge, MA: Harvard University Press.

Mail & Guardian. (2011) '"Aerotropolis" to boost economy', Mail & Guardian, [online] 28 October, available from: https://mg.co.za/article/2011-10-28-aerotropolis-to-boost-economy/

Masango, R., Mfene, P. and Henna, T. (2013) 'An analysis of factors that negatively affect the performance of ward committees in the Buffalo City Municipality', *Africa Insight*, 43(1): 91–104.

McFarlane, C. and Silver, J. (2017) 'The poolitical city: "Seeing sanitation" and making the urban political in Cape Town', *Antipode*, 49(1): 125–48.

Milan, S. and Gutiérrez, M. (2015) 'Citizens' media meets big data: The emergence of data activism', *Mediaciones*, 11(4): 120–33.

Milan, S. and Treré, E. (2019) 'Big data from the South(s): Beyond data universalism', *Television & New Media*, 20(4): 319–35.

Mitchell, H. and Odendaal, N. (2015) 'From the fringes: South Africa's smart township citizens', in M. Foth, M. Brynskov and T. Ojala (eds) *Citizen's Right to the Digital City: Urban Interfaces, Activism, and Placemaking*, Singapore: Springer, pp 137–59.

Moeng, K. (2011) reporting the words of Mondli Gungubele in *SowetanLive* (14 April 2011), available from: https://www.sowetanlive.co.za/news/2011-04-14-mondli-gungubele-wants-the-region-to-be-the-economic-hub-of-south-africa/

Monahan, T. (2005) *Globalization, Technological Change, and Public Education,* London: Routledge.

Ncoyini, S.S. and Cilliers, L. (2020) 'Factors that influence knowledge management systems to improve knowledge transfer in local government: A case study of Buffalo City Metropolitan Municipality, Eastern Cape, South Africa', *SA Journal of Human Resource Management,* 18. https://doi.org/10.4102/sajhrm.v18i0.1147

Odendaal, N. (2021) 'Everyday urbanisms and the importance of place: Exploring the elements of the emancipatory smart city', *Urban Studies,* 58(3): 639–54.

Odendaal, N. and McCann, A. (2016) 'Spatial planning in the global South: Reflections on the Cape Town spatial development framework', *International Development Planning Review,* 38(4): 405–23.

Pillay, C. (2019) 'Inside OpenUp', Medium, [online] 6 September, available from: https://medium.com/openup/inside-openup-64c30a83f751

Planact. (2019) Intervention photograph.

Planact. (2021) 'About Planact', Planact, [online], available from: https://planact.org.za/about-us/

Pollio, A. (2022) 'Acceleration, development and technocapitalism at the Silicon Cape of Africa', *Economy and Society,* 51(1): 46–70.

Power, M. (2022) 'Theorizing the economy of traces: From audit society to surveillance capitalism', *Organization Theory,* 3(3). https://doi.org/10.1177/26317877211052296

Ricker, B., Cinnamon, J. and Dierwechter, Y. (2020) 'When open data and data activism meet: An analysis of civic participation in Cape Town, South Africa', *The Canadian Geographer/Le Géographe canadien,* 64: 359–73.

Republic of South Africa (RSA). (1996) Constitution of the Republic of South Africa, Pretoria: Government Printing Works RSA.

Republic of South Africa (RSA). (2000) Municipal Systems Act, Pretoria: Government Printing Works RSA.

Republic of South Africa (RSA). (2013) National Development Plan, Pretoria: Government Printing Works RSA.

Republic of South Africa (RSA). (2016) Integrated Urban Development Framework, Pretoria: Government Printing Works RSA.

Ruppert, E., Isin, E. and Bigo, D. (2017) 'Data politics', *Big Data & Society,* 4(2). https://doi.org/10.1177/2053951717717749

Sikhakane, B.H. and Reddy, P.S. (2009) 'Local government restructuring and transformation in South Africa with specific reference to challenges faced by Buffalo City Municipality', *Administratio Publica,* 17(4): 232–51.

Söderström, O. (2018) 'Analysing urban government at a distance: With and beyond actor-network theory', in K. Kurath, J. Ruegg, J. Paulos and M. Marskamp (eds) *Relational Planning: Tracing Artefacts, Agency and Practices*, Cham: Palgrave Macmillan, pp 29–50.

Söderström, O., Blake, E. and Odendaal, N. (2021) 'More-than-local, more-than-mobile: The smart city effect in South Africa', *Geoforum*, 122: 103–17.

Streule, M., Karaman, O., Sawyer, L. and Schmid, C. (2020) 'Popular urbanization: Conceptualizing urbanization processes beyond informality', *International Journal of Urban and Regional Research*, 44(4): 652–72.

van Dijck, J. (2014) 'Datafication, dataism and dataveillance: Big Data between scientific paradigm and ideology, *Surveillance & Society*, 12(2): 197–208.

VPUU. (2021) 'Who we are', VPUU, [online], available from: http://vpuu.org.za/who-we-are/

12

Epilogue: Beyond Data and Crisis

Orit Halpern

If data have politics, then what form of politics are these? This book, *Data Power in Action,* sets out to ask precisely this question. The stakes could not be higher – the urban is a key site for asking about the future of politics, but also of human and more-than-human habitat and life in the future.

In Western genealogies, deriving from Greece, the *polis* is the city-state. The term *polis* denotes both a place to inhabit (*oikos)* but also a place of power and politics (the state). Cities were also the place of the *demos*, the site where the people came to be seen to power and also make claims *of and for power.*

Moreover, the *polis* demarcated the territory separating culture from nature. The animals could not be political beasts. The city, therefore, is the birthplace in the Western myth of politics, and the place where History happens, if we understand history as the realm of only human actions. These were not all people – only citizens had power, and most of the population were not citizens, but it was the fantasized location for democracy. The city, at least in some narratives, was the habitat for both the human and the political, particularly democratic politics.

I open with these highly problematic and very geographically and culturally specific narratives of the *polis* to highlight a fundamental question that appears to occupy dominant discussions of machine learning, AI, and big data in our present. By definition it would appear that the drive for data-driven decision-making heralds a breakdown of older Western models of decision-making and human agency. If humans need big data to make decisions, and increasingly *automated* decisions, then how should we understand the possibility and problems that such a condition heralds for polity, politics, and the urban in our present? This book clearly marks out a map of some directions by which we might pursue this question, and more importantly, its stakes. For ultimately, the question of data, urbanism, and technology is one of how do we wish to live in the future? And who even constitutes this 'we'? What polities (and powers) will be emerging? And what futures are we imagining through and with our technologies?

The technosphere

The language of crisis by its very definition already suggests disruptions to norms, questions of survival, and critiques of progress. However, the contemporary situation of global warming, environmental degradation, political economic transition, and technological change appears to turn this crisis situation into a permanent condition *and* technological opportunity. Every crisis is a reason for more computing! This relationship between crisis, technology, and automating decision-making (or at least imagining doing so) has subsequent bearing on how we understand human agency and political action.

In fact, the idea that we depend on our machines for our survival has now taken the shape of evolutionary imperative. For example, in 2014 the geologist Peter K. Haff, a member of the working group within the International Commission on Stratigraphy coined a term with considerable bearing on this relationship between habitat and politics: 'technosphere'. He argued:

> the set of large-scale networked technologies that underlie and make possible rapid extraction from the Earth of large quantities of free energy and subsequent power generation, long-distance, nearly instantaneous communication, rapid long- distance energy and mass transport, the existence and operation of modern governmental and other bureaucracies, high-intensity industrial and manufacturing operations including regional, continental and global distribution of food and other goods, and a myriad additional 'artificial' or 'non-natural' processes without which modern civilization and its present 7×10^9 human constituents could not exist.
>
> (Haff, 2014b, 301–2)

For Haff, the term 'technosphere' is a cyborgian sphere. Within it, humans are not ascendant, but rather part of networks of technology that are necessary to sustain life.

From a historical perspective, what is interesting is that Haff proposes this term to replace an older idea of the biosphere. Life for Haff is fundamentally dependent on technology. Humans can no longer understand themselves as separable or independent from technical systems. This comprehension was already anticipated by feminists such as Donna Haraway who imagined the intimate dependencies between humans, machines, and other life forms as offering the potential for new forms of life and violence (Haraway, 1991).

The technosphere, however, denotes a certain survivalist understanding of technology. The term emerged within the context of a broader debate over the Anthropocene, providing a necessary corrective to overly anthropocentric accounts of climate change by 'suggest[ing] a more detached view of an emerging geological process that has entrained humans as essential components that support its dynamics' (Haff, 2014b, 302). For Haff, humans are '"parts" of the technosphere – subcomponents essential for system function' (Haff, 2014b, 306). As Haff notes, the vast majority of humans currently living on the earth rely on the technosphere not simply for cell phones and computers, but also for 'essentials of life such as food and water, which, for the billions of humans alive today, are available only as a consequence of the function of the technosphere (e.g. fertilizers, mechanized farm equipment, housing, efficient long-distance transportation, pesticides, medicines and so on)' (Haff, 2014a, 133). Haff stresses that though the agency of the technosphere is not the same as human agency, we cannot truly address problems such as global warming or the future of urbanization unless we 'recognize and engage that agency [of the technosphere]'(Haff, 2014b).

What Haff suggests is that we no longer inhabit cities or urban spaces but rather a new sphere – the technosphere. This sphere is necessary for our survival. New terms such as 'planetary urbanization' repeat this idea. There is nowhere, advocates of this view argue, that is not mechanized, computerized, and financialized left on earth (even the rural, wilderness, and hinterlands must be subsumed to the urban) and the future, therefore, must be decided through the management of this urbanity (Brenner and Schmid, 2011).

This planetary urban or technical environment also comes with a demand, even or especially from critics, to rethink the human and human agency. In 'The climate of history' postcolonial historian Dipesh Chakrabarty extends Haff's suggestion to reframe human agency and history. He suggests the concept of 'the planetary,' to replace older ideas of the global or history. Planetarity, Chakrabarty (2009) argues via Haff, in invoking planetary sciences ties the human as a cultural/historical agent to the geological, biological, and 'scientific' being (scientific being the subject of the human sciences such as demography and economics):

> The planet puts us in the same position as any other creature. Our creaturely life, collectively considered, is our animal life as a species, a life that, pace Kant, humans cannot ever altogether escape. Our encounter with the planet in humanist thought opens up a conceptual space for the emergence of a possible philosophical anthropology that will be able to think capitalism and our species life together, from both within and against our immediate human concerns and aspirations.
>
> (Chakrabarty, 2009, 29)

While there are clear limits to this analysis, it begins to suggest that our current situation demands an account that is more than human, and can traverse the human, technical, and 'natural'. The planetary, therefore, denotes replacement or hybridization of the social with the technical. Human beings, however, have only become geological forces by means of specific kinds of technologies, and those technologies are fundamentally in all these accounts those of 'networks', 'instantaneous communication', 'modern organizations', and forms of derivation and extraction that exceed the industrial limitation of labour power. Almost without question, these authors all assume a new sphere whose main evolutionary driving force is, although only suggested, is as media theorist Benjamin Bratton puts it, 'planetary scale computation' (Bratton, 2016).

There is a subtle but important element at work here in these discourses of planetary urbanization and the technosphere that I want to summarize. First, in these narratives, even postcolonial stories of the Anthropocene, technology seems to have replaced the biosphere at least since the early 2000s as the imagined zone of earth that makes life possible. In being the necessary substrate for survival, the technosphere and by extension, the urban, has been recast as both the greatest threat to human (and other species) survival and the only route to salvation. In a subtle but really important move, where the geological, biological, and physical sciences are made commensurate with the historical, cultural, and social spheres we also see the shift from languages of progress, planning, and utopia to those of adaptation, evolution, extinction, and populations (or big data). The technosphere has taken over the discourses of survival now framed in species and languages of evolution. In urban planning, ideas like planetary urbanization assume the technosphere is *the* condition of our present and that all places can and *must* become urban to survive: a sort of new imperative to colonize all life into the city. Such a possibility offers both frightening frontiers for colonial expansions of technology, power, and capital (belt roads, smart cities, extreme infrastructures, even post-planetary extraction, and so forth), but also perhaps new avenues for expansions of the political.

Determinism and reason

The problem with such Anthropocenic understandings of the technosphere is the underlying concept of determinism. As Prince K. Guma (Chapter 9) notes, 'studies tend to take a determinist view that does not grant urban stakeholders outside the corporate framework much agency in shaping digital presents and futures'.

Bio-determinism has a long and rather sordid history affiliated with racism and eugenics. Technical determinism may be no better. Discourses of planetary urbanization, the technosphere, and smartness, all marry technology to species survival at planetary scales, recasting a biological and evolutionary mandate on discourses of data, AI, and ubiquitous computing (Halpern, 2019). This logic substantiates itself on discourses of crisis and catastrophe, of which pandemics and climate change are central. As Datta and Söderström powerfully demonstrate here, in Varanasi, India, COVID-19 provided the 'crisis' permitting the test-bedding, experimentation, and expansion of data-driven systems. It is perhaps the fact that the expansion and roll-outs of these techniques were never seamless or perfect that even provided justification for ongoing experimentation and expansion, systems are never perfect but subjective (as the introduction makes very clear).

Smartness, understood in terms of adaptive data-driven systems that can change and 'feed back' or respond to their environments thus become 'mandatory' for survival *and* evolution. Without these techniques and technologies of smartness adaptation and change become unthinkable. Just as smart cities are necessary to deal with the crises of COVID, climate change, and so forth (clearly depicted in this book).

For advocates of such schemes, smartness and the need to collect data thus come to have the force and irresistibility of a law of nature. It is a form of 'smartmentality', to use Ying Xu, Federico Caprotti, and Shiuh-Shen Chien's framing in Chapter 8, that is seemingly inevitable and naturalized, serving different ends depending on context. All social process 'must' become smart. The smartness mandate seems to be a mandate in part because of the high stakes involved: for its advocates, we *must* become smart or else go extinct as a species. A system only attains 'smartness' when it achieves the capacity to adjust to any new and unexpected threats and possibilities that may emerge from the city's ecological, political, social, and economic environments (a capacity that is generally referred to in planning documents with the term 'resilience'). In short, a smart city is a site of perpetual *learning*, and a city is smart when it achieves the capacity to engage in perpetual learning.

Such logics enjoin us to smartness – rather than, for example, rationality – in order to underscore the inability of unassisted human reason to understand and cope with the modern challenges that humans face; as a consequence of this incapacity, humans need learning processes that take place largely within computer systems.

Smartness versus reason

However, it is precisely that smartness is not reason that such concepts might offer some new imaginaries for thinking both urbanism and the politics of big data. I emphasize the term rationality here because smartness is not reason as understood since between the 17th and 19th centuries and related to the Enlightenment and later liberalism. A central tenet to artificial intelligence and to smartness (at least in certain places and particularly within neoliberalism) is that unassisted human reason cannot fully understand and cope with the modern challenges that humans face, and that, as a consequence, humans need learning processes that take place at least partly within or with computer systems. This notion of the fallibility and subjectivity of the human is repeated in discussions of cyborgs, technospheres, and the Anthropocene.

This notion that decision-making and intelligence might be subjective, networked, and decentered is, in fact, one of the central ideas shared between economics, computer science, and psychology since the 1950s (Erickson et al, 2013; Halpern and Mitchell, 2023). Some of the same developments in computer science, psychology, and evolutionary theory inspired, for example, neoliberal economic theorists such as F.A. Hayek and architects of computer-learning processes such as Frank Rosenblatt (and Rosenblatt in turn drew on Hayek). Both neoliberalism and smartness also share a similar commitment to 'epistemological modesty,' in the sense that they share a common assumption: since no single individual or group of individuals can predict what the future will bring, one must rely on structures and mechanisms that perpetually adjust and 'learn' (Halpern, 2022).

For neoliberals, this structure is 'the market'; for advocates of smartness, it is smart technologies and processes. The relationship between the neoliberal market and smartness can sometimes be more than simply analogy or resonance, since smart technologies often rely on prices as a way of assigning 'weights' in learning algorithms. These weights enable an algorithm to adjust its predictions overtime, and this is an essential part of machine learning. However, smartness is not reducible to neoliberalism, a point made throughout the text *Data Power in Action*. As case study after case study reveal in Nairobi, Varanasi, London, and many other cities, smartness opens new avenues for activity that is not always reducible or directed into financial markets.

This opens to a new question of how we might reimagine the relationship between technology and nature through smartness without becoming neo-Darwinian or neoliberal. Most critically, how might we contend with determinism, and the assumptions that certain paths of technical development are inevitable.

Smartness is not omniscience

While smartness, and more broadly technical determinism, would gesture to a language of norms and determinism – who, after all wants to live in a dumb city? And who would not want a smarter city? – it is possible, and certainly this volume suggests just this, to understand smartness otherwise. This is one of the opening premises of this volume, as Datta and Söderström argue, 'data politics emerges at this junction as the data deluge becomes highly diversified, personalized, compartmentalized, but also fragmented, disconnected, and uneven' (Chapter 1). It is precisely because data have to be captured, or are *capta*, to cite Rob Kitchin, that data is also not omniscient, objective, or disembodied. Rather, it is embodied, contextualized, situated, and ultimately made through social forces. This penetration of data into life is what makes it possible to link data to both crisis and politics following the logic of this text.

If one takes the subjective nature of data as the foundational premise of smartness and big data analytics, then one might argue that the opposite of smartness might be said to be omniscience.[1] Put otherwise, smartness's opposite is not a normative characteristic of human bodies and subjects, such as 'dumbness' or 'brilliance' or 'genius', but rather absolute control and objectivity. At first, this would appear unintuitive. After all, are not data-driven systems there to permit total control and surveillance by a select few? This is certainly possible. But the language of smartness suggests that we can either aspire to omniscience – which, if attainable, would indeed allow us to take all contingencies into account beforehand – or we can recognize that omniscience is impossible for mere mortals, and instead aim for smartness, which means perpetual learning in the light of changing circumstances through and with other humans, animals, and machines. As Rob Kitchin and AbdouMaliq Simone note in Chapters 2 and 5, data are always subjective and 'made', offering new forms of relations and producing new types of life. Simone suggests that we must go 'beyond conventional modes of calculation, measurement, and value' (Chapter 5). This 'beyond' suggests that both critiques and proponents of big data systems always recognize the failures and inadequacies of existing data and modes of calculation. Smartness is very difficult to critique for precisely this reason. Any bias, any limitation to measurement, and any limit to the predictive analytics is not a reason to disband systems, but a call to improve them. Add more sensors! Increase technological interventions! Ergo the link Datta and Söderström insist on in the framing of this text: data and crisis beget each other. Failures to predict and avoid crisis are reasons to increase computational power and data collection. This mode of turning failures into frontiers for technical manipulation and enhancement is central to the logics of big data, platforms, and smart infrastructures. As Simone also suggests, this productivity of limits is also a possible site of imagination and emergence – since the regular failure of systems also is a call to envision new ways to measure, model, and make the world.

Thus, despite its heavy reliance on cutting-edge technologies, smartness can be opposed to technocratic visions of social change (of course, not always). This is to say that smartness, in some scenarios but not all (China, for example), is not necessarily a call for centralized planning. In fact, smartness often assumes a decentered, self-organizing, or technophilic form of control not invested in civil servants or organized planning but rather in engineering. This should never be understood as apolitical or without power, however.

Politics is simply happening in the design and management of computational systems. In this manner, new agents and relations are regularly being created. The analysis of smartness and big data platforms can contest, for example, the technocratic distinction between experts and non-experts in favour of the claim that everyone has knowledge to contribute. In capital accumulation, as Petter Törnberg notes:

> The platform mode of regulation implies a new way of seeing, as platforms see those governed through the novel epistemology of Big Data – cluster-based, bottom-up, and relational – and exerting control through the design of programmable social infrastructures. This signifies a fundamental shift in the regime of power, lying at the heart of the societal transformations emerging from digitalization and platformization.
>
> (Chapter 3)

Big data infrastructures and platforms thus permit new forms of governing that is 'cluster based, bottom-up, and relational', a form of economy and government that creates new sites of value around control and data.

This is an extractive and self-referential form of economy, according to Törnberg. But such systems also pose other potentials. On the productive possibilities of such smart systems, the DIY ethos demonstrated in the work by Powell, Cinnamon, and Guma all demonstrate the dynamic and emergent properties of learning collectively (for better or worse) on platforms. In this sense, it can also become an appeal to include previously marginalized voices, and it demands that everyone, including those who are privileged, become perpetual learners.

Activating such potential in and through platforms and smart infrastructures –that no one is omniscient and that the marginalized may also make data – is a significant element of what politics has become in the account laid out by the many authors engaging data and crisis. From this perspective, the basic idea of reflexive big data systems and smartness, even if some of its current implementations may be considered problematic, offers an alternative to command, control, and communication paradigms or to reductive understandings of surveillance capital (Zuboff, 2019). The importance of smartness and big data-driven systems as contested categories and practices is the central imperative of this text. The authors all urge us to situate and contest deterministic ideologies of data in order to find new paths towards justice and agency.

Perpetual learning

If there is a site of potential in 'smartness' then it might lie in the possibility of learning. The inhabitants of a smart city are, in theory, also perpetual learners. As the stories of survival and platforms (Guma, Chapter 9), data and action (Powell, Chapter 4) and contingency (Simone, Chapter 5) suggest, the denizens of cities are also always manipulating, hacking, reinventing, and learning from, with, and around the technologies they must negotiate. As the smart city constantly adapts, the people who live in it will also have to adjust. In theory, this is part of the logic. When I once asked Cisco engineers in 2013 at a large greenfield development in South Korea – Songdo – how their model might be transported elsewhere, they responded that the smart city's smartness is not supposed to be imposed on its urban inhabitants from above; rather, this smartness is supposed to result from the combination of the inhabitants' unique individual perspectives and choices. While the method (computer-assisted data-driven decision-making) might be the same, the forms and decisions would be decided by context. Smartness presumes that these acts of combination cannot be accomplished by humans alone, but require the assistance of computing processes – and, more specifically, of algorithms that teach the smart city (and its inhabitants) new ways in which to learn. Very much like 'the market' of neoliberal economic theory, smartness optimizes processes by combining multiple perspectives in a way that cannot be achieved by any group of human planners. For some of its advocates, the ability of smartness to automate the combination of an enormous number of individual perspectives makes it possible to imagine that one could perhaps replace politics – that messy realm of self-interest, which often only seems fully open to a select few – with technological processes that could actually achieve what democracy only promises.

However, learning is also a possible route to alternative futures. The importance of smartness – as ideology, as an ever-changing set of technologies and techniques, but also as a possible focal point for hope – becomes especially evident when viewed from the perspective of the Anthropocene, from which I began this discussion. This crisis includes global warming, the increasing dominance of one-crop agriculture, pandemics, wealth disparities, and a plethora of other global dangers. It seems clear to us that humanity has arrived at this point as a result of capitalism, and it seems equally evident that capitalism itself cannot fix this problem, no matter how many innovative new forms of market its advocates may come up with (carbon offset markets, new forms of insurance for endangered coastal areas, and so on). However, precisely because it is not identical to neoliberalism, smartness retains its potential within this context. For example, Winona LaDuke and Deborah Cowen (2020) introduce ideas of 'alimentary infrastructure' as a way to think smart energy infrastructures away from the market-based principles of contemporary smart electrical grids and towards indigenous calls for environmental custodianship and sovereignty. And in this volume, there are multiple examinations of data as both 'weaponized and democratized' in urban contexts (Chapter 1). Powell (Chapter 4) and Cinnamon (Chapter 10) note that scale is a critical feature to data politics, and a site for action. As Powell notes:

> Between these two scales [the planetary and the local or embodied subjectivities of city inhabitants] lies the potential to investigate ethics as practice as a means to surface other forms of knowledge and care that might be necessary for a flourishing existence in a state of perpetual crisis. These practices might include research practices like data walking, or experiments in creating multiscalar relational structures that allow for dynamics of mutual aid and support to proliferate.
>
> (Chapter 4)

For Powell, there are new forms of dispossession by corporate actors banking data for the UK and London City (such as Palantir), but there are also new forms of action and perhaps infrastructures of care through 'data-making' (a key term for this text).

In my own work, Mitchell and I introduced the idea of *the biopolitical learning consensus*. Hardly sexy or great like smartness, we still thought it gestured to some of the possibilities of thinking in terms of populations and through data might offer. We also thought to insist on linking our analysis to updating and interrogating older questions of power, governmentality, and subjectivity that such biopolitics affords. Such a learning consensus agrees that unassisted human reason cannot fully understand and cope with the modern challenges that humans face, and that, as a consequence, humans need learning processes that take place at least partly within computer systems. Yet for the biopolitical learning consensus, the limits on unassisted human reason stem less from intrinsic limits of rationality than from the fact that there is no one set of humans who would ever be able to define the nature and contents of 'human reason.' Human rationality is not something that can be defined and axiomized by, for example, game theorists of the 1960s; rather, it is a capacity that is always at work, and at work differently, in every human collective. Learning processes that take place in part within computer systems thus need to remain open to different perspectives on both what constitutes rationality and what counts as learning.

In place of a mandate to be smart, the biopolitical learning consensus seeks to bring these different perspectives together; that is, to feel and think together-with. And unlike past examples of 'consensus' that in fact represented agreements among a very small number of people – for example, the neoliberal 'Washington Consensus' that emerged in the late 1980s – this consensus has no geographic location, but rather comes together through the distributed efforts of those interested in learning in all its forms.

The question of what 'learning' does and can mean is clearly a key question for any effort to bring smartness and democracy closer to each other. If smart learning is itself based on a Darwinian evolutionary image of winners ruthlessly culled from losers, smartness will always find itself unable to escape the ever-extendable horizon of the market. This book has, however, reminded us that there are many forms of learning, and that in different locales and spaces smartness and big data infrastructures take different and often contested instantiations.

This is a point forcefully made throughout this volume. Collectively this work critiques the inevitability narratives of ICT and data-driven urban development in Kenya, South Africa, and India. But learning could also be understood to be about narratives and storytelling. The introduction of this text posits these strategies as critical. In fact, Datta and Söderström argue that 'storytelling, animation, and seamfulness' are their central concerns. How do we learn about the city? How do we narrate injustice and violence so that they become visible and enable grassroots activists 'to disrupt "global city" narratives authored by governments and those in power' (Cinnamon, Chapter 10). And how are violence and injustice also hidden, how is labour made invisible? These are questions of learning, aesthetics, and practice. If we can learn differently, which also demands new modes of representation and storytelling about our world, perhaps we can also change, not through rejection of big data and technology but through envisioning new assemblages, practices, and aesthetics *using* data.

The first question in the possibility of cyborgness we might remember is that the cyborg is a 'chimera', a myth, and therefore is not only a reality, but also a desire and an imaginary but only when separated from discourses of existential threat. Pointing out the link between crisis and data does just this.

This book on data and crisis thus calls our attention to how we have naturalized the idea that the coming catastrophe mandates technological intervention. It also moves our discourse away from absolute notions of technical dependence and evolutionary determinism now recast from the biological to the technological, but rather sites our concerns on *practice*. The editors state this at the start: how does this naturalization happen? In essence, the authors in this text force us to consider another term: *power*. Not only when power is legible and representable in politics, but the tactics, techniques, epistemologies, materialities that reorganize different agents into new assemblages. This cannot be overstated, since whether we label our contemporary condition Anthropocene, the Capitalocene, or any other term, all the terms continue to exert notions of agency and determinism that perhaps do not do due justice to the complexity of our contemporary planetary conditions.

I take the book as a document of the fact that no one site or researcher can comprehend planetary conditions of technology or imagine futurity. We might reflexively take our own research as a reminder that there are experiences that can only emerge through the global networks of sensory and measuring instrumentations. There are, in this sense, radical possibilities in realizing that learning and experience might not be internal to subjects but also shared. Perhaps these are just realizations of what we have known all along: that our worlds are comprised of relationships to others. Every market crash, every moment platforms are hacked or manipulated, every time systems do not do what they are intended to, reveals our fundamental and absolute technical co-dependence and relationality to one another. Our machines regularly remind us what we always need to remember: that all systems are programmed and, therefore, changeable.

Note

[1] I am indebted to my co-author on *The Smartness Mandate*, Robert Mitchell, for these ideas.

References

Bratton, B.H. (2016) *The Stack: On Software and Sovereignty*, Cambridge, MA: MIT Press.

Brenner, N. and Schmid, C. (2011) 'Planetary urbanization', in M. Gandy (ed) *Urban Constellations*, Berlin: Jovis Verlag, pp 10–13.

Chakrabarty, D. (2009) 'The climate of history: Four theses', *Critical Inquiry*, 35(2): 197–222.

Erickson, P., Klein, J.L., Daston, L., Lemov, R., Sturm, T. and Gordin, M.D. (2013) *How Reason Almost Lost Its Mind: The Strange Career of Cold War Rationality*, Chicago: University of Chicago Press.

Haff, P.K. (2014a) 'Humans and technology in the Anthropocene: Six rules', *The Anthropocene Review*, 1(2): 126–36.

Haff, P.K. (2014b) 'Technology as a geological phenomenon: Implications for human well-being', in C.N. Waters, J.A. Zalasiewicz, M. Williams, M.A. Ellis and A.M. Snelling (eds) *A Stratigraphical Basis for the Anthropocene*, London: Geological Society, pp 301–9.

Halpern, O. (2019) 'Hopeful resilience', in L. Kurgan and D. Brawley (eds) *Ways of Knowing Cities*, New York: Columbia Books on Architecture and the City, pp 23–34.

Halpern, O. (2022) 'The future will not be calculated: Neural nets, neo-liberalism, and reactionary politics', *Critical Inquiry*, 48(2): 334–59.

Halpern, O. and Mitchell, R. (2023) *The Smartness Mandate*, Cambridge: MIT Press.

Haraway, D.J. (1991) 'A cyborg manifesto: Science, technology, and socialist-feminism in the late twentieth century', in D.J. Haraway, *Simians, Cyborgs, and Women: The Reinvention of Nature*, New York: Routledge, pp 149–81.

LaDuke, W. and Cowen, D. (2020) 'Beyond Wiindigo infrastructure', *South Atlantic Quarterly*, 119(2): 243–65.

Zuboff, S. (2019) *The Age of Surveillance Capitalism: The Fight for a Human Future at the New Frontier of Power*, New York: Hachette Book Group.

Index

References to figures appear in *italics*. References to endnotes
show the page number and the note number (140n1).

Printed and bound by CPI Group (UK) Ltd, Croydon, CR0 4YY

27/10/2024

14580559-0003